The Semiconductor Business

The Economics of Technological Change
Edwin Mansfield, *General Editor*

The Semiconductor Business
The Economics of Rapid Growth and Decline

Franco Malerba

The University of Wisconsin Press

Published 1985 in the United States of America by
The University of Wisconsin Press
114 North Murray Street
Madison, Wisconsin 53715

Published in Great Britain by
Frances Pinter (Publishers) Limited
25 Floral Street, London WC2E 9DS

First printing
Printed in Great Britain

Library of Congress Cataloging-in-Publication Data

Malerba, Franco, 1950–
    The semiconductor business.
    Bibliography: p.
    Includes index.
    1. Semiconductor industry — Europe.
    2. Semiconductor industry — United States.
    3. Semiconductor industry — Japan. I. Title.
HD 9696.S43E8555   1985   338.4'762138152 85-40372
ISBN 0-299-10460-5

Typeset by Joshua Associates, Oxford
Printed by SRP Ltd, Exeter

Contents

For Pamela

Foreword

This study should be read in two different ways. It is, on the one hand, a detailed analysis of the evolution of the European semiconductor industry, and a comparison of European developments with those in the United States and Japan. But it is also a contribution to the understanding of technical change in industry, and of public policy bearing on technical change.

The book is directed toward a specific set of questions. Few analysts would argue that, prior to the Second World War, the American electronics industry was technologically more advanced than the German or the British. Yet today, Europe's semiconductor and computer industries clearly lag behind the American and the Japanese. How did this come about? Malerba assigns a good deal of the credit to the massive and adventuresome American defence and space programs which provided both a lucrative market and sources of funding for American firms who set their technological sights high. Malerba also points out the major role played by the new American firms in the technological history of semiconductors, and gives credit to the openness of the American semiconductor industry, and the willingness of defence contractors to work with new firms. While other authors have argued similarly, Malerba's detailed analysis of why these features of the American scene were important, and why the Europeans were unable to duplicate or offset these advantages, significantly advances the discussions.

Malerba argues that by the mid-1960s the American defence and space programs were no longer playing their earlier crucial catalytic role, but that, by that time, the American firms had a lead over the Europeans that has proved difficult to make up. Somehow the Japanese, on the other hand, were able to catch up in many important areas. Malerba's discussion of why they were able to do so, and how the Japanese context differed from the European, is interesting. Part of the answer is that the Japanese were able to protect their large home market from both American exports and American direct investments. But there is much more to Malerba's story than that.

While focused on a particular industry, Malerba's study is a significant contribution to general understanding of the processes of technical advance in industry. The last decade has seen a renaissance of Schumpeterian ideas in economics, and the sharpening, modification and elaboration of these as a result of a significant amount of empirical and theoretical research. Malerba's study has been guided by this burgeoning new literature and is a major addition to it. The new literature has defined a class of technologies as 'cumulative', in the sense that each advance seems to lay the basis for the next, and Malerba clearly identifies semiconductor technology as belonging to this class. Theoretical work has suggested that, in such technologies, a head start can be decisive in even a long race. Malerba's study indicates that this has indeed been the case in semiconductors. Malerba's study also contributes to evolving understanding of the relationships between industrial structure and technological advance. In particular, Malerba has many interesting things to say about the different kinds of R&D incentives, and constraints, faced, on the one hand, by vertically integrated companies producing final products which use semiconductors, as well as semiconductors themselves, and, on the other, by companies that specialize in the production of semiconductor devices for sale. There are good reasons to believe that the relationships Malerba observes in the semiconductor industry obtain also in other industries.

While technological change in the semiconductor industry shows many patterns similar to those in other industries, semiconductors also are rather special in certain respects. One of these is that, arguably, it has been the advance of semiconductor technology, more than any other technological development, that, since 1960, has led and shaped the pattern of technological advance more broadly. Thus, understanding this particular story is especially important. A second, but related, characteristic of the semiconductor field is the striking attention governments have paid to that industry in recent years, under the premise that only if a nation's semiconductor industry is strong, will a nation's manufacturing industry be strong generally. Malerba's book tells the story of the rise of national policies in support of the national European semiconductor industries.

And yet, there is cause to question the basic premise. The period when American firms were drawing away from European ones in semiconductor capabilities also was one of very rapid growth and high employment. Japan's post-war growth surge began long before she acquired competence in semiconductors.

There is also the question of whether, in this day and age, national capabilities in an industry like semiconductors have any clear meaning. As Malerba describes it, the major companies in the semiconductor industry are by now mostly multi-nationals, with plants and R&D facilities located in a variety of different countries. Joint ventures among companies in different countries are increasingly common. These developments are occurring at the same time as the governments of individual nations are increasing their efforts to define and bolster their national industries. It will be interesting to see how these opposing currents work out.

Richard R. Nelson
April 1985   Yale University

Preface

This book is about technological innovation, dynamic competition and evolution in a high-technology industry: the semiconductor industry. The semiconductor industry is an interesting case for such a study for several reasons. It has been one of the fastest growing and most innovative industries throughout the post-war period. Firms within this industry, moreover, have had to cope with constantly changing technologies, and have experienced constantly changing competitive advantages as well as lead and lag times. The semiconductor industry, therefore, best exemplifies an industry in which the Schumpeterian 'perennial gale of creative destruction' (innovation) has played a major role.

The history of the semiconductor industry has been characterized by the highly innovative and competitive position of the American industry and the success of the Japanese industry to 'catch up'. Between these two cases lies a major puzzle about the European industry: why, despite its success during the early history of the industry, did the European industry fall behind and why, later, was it unable to catch up with its American counterpart? This book attempts to solve this puzzle. The approach used is a dynamic, evolutionary and historical one, framed in a comparative perspective.

This book has benefited greatly from the contributions of several people. It originated as a Ph.D. thesis written at Yale University under the guidance of Professors Richard Nelson and Richard Levin. To Richard Nelson I owe a great intellectual and personal debt; he has had a profound influence over my conceptions about technological change and about the evolution of industries. Throughout the course of this research, moreover, he has been helpful and encouraging, as well as insightful and critical. Richard Levin has also provided invaluable suggestions and stimulus during the various stages of this work. I am also grateful for the helpful comments of Professors Sidney Winter and Merton Peck of Yale University; Edwin Mansfield of the University of Pennsylvania; David Mowery of Carnegie Mellon University and Stuart MacDonald of the University of Queensland.

Finally, I would like to thank the late Professor Innocenzo Gasparini of Bocconi University for his enthusiastic support and advice and Professor Fabrizio Onida of Bocconi University for his contributions to my understanding of the causes of the international competitiveness of industrialized countries.

A large part of the material and data on which this analysis was built derives from interviews with industry executives. A full list would be too long. However, some deserve special acknowledgement and thanks for their help and assistance: Dr Hogan of Fairchild; Dr Parker of Intel; Mr Volkholz of Signetics; Drs Edlinger, Hazewinds, Huart, Van Iersel, Wel, Esseling, Manintveld and Speyer of Philips; Drs Hofmeister, Richard, Wiesner, Schoen and Bohle of Siemens; Mr Bohmardt,

Drs Dahlberg, Ehlbeck and Schneider of AEG-Telefunken; Mr Karnatzky of Intermetall; Mr Gauzhorn of IBM; Dr Dugas of Thomson-CSF; Dr Shepherd of Ferranti; Dr Larkin of Plessey; Mr Borel of EFCIS; Drs Paletto, Villa, Gradi, Zocchi, Borri, Riva, Murari, Faini, Seragnoli and Cabarè of SGS-ATES; Drs Lamborghini, Gandi, Toppi and Conti of Olivetti; Ing. Costamagna and Ricciardi of Alfa Romeo; Ing. Alderisio of IMI; Dr Orlando of Honeywell; Prof. Fabri of Italtel; Ing. Muratori of STET.

In addition, the following experts and government officials have also provided valuable contributions: Dr Klesken; Dr Hinkelman of SIA; Drs Bernecker and Rembser of the BMFT; Dr Scholz of IFO; Drs Pavitt, Dosi and Sciberras of SPRU; Dr McLean of *Electronic Times*; Mr Noyce of the Department of Trade and Industry; Dr Marr of NEDO; Drs Truel, Delapierre and Cohen; Drs Bizec and Pollard of CNET; Dr Robert of DIELI; Ing. Morganti and Dr DeBrabandt of RESEAU, and Prof. Momigliano.

During the writing of this book, I have also greatly benefited from the support of Yale University and Bocconi University, the Francesco Ferrara Fellowships, the Sloan Foundation, the US National Science Foundation and the Italian Consiglio Nazionale delle Ricerche. I want also to thank Peter Moulson of Frances Pinter and Gordon Lester-Massman of the University of Wisconsin Press. It has been a pleasure to work with these editors on both sides of the Atlantic.

Finally, I owe the greatest debt to Pamela Adams Malerba. She provided constructive criticism, valuable suggestions and editorial help, and has supported the work with enthusiasm, understanding, warmth and intelligence. To her I dedicate this book.

1 Why did the European industry decline and why was it then unable to catch up?

This book is concerned with a major puzzle about the European semiconductor industry: why did the European industry decline from a position equal to that of the American industry and superior to that of the Japanese industry during the 1950s and why was it unable to catch up with the American industry later on as the Japanese industry did? This puzzle is highlighted by the history of the world semiconductor industry. Prior to World War II, European firms, and particularly German firms, were at the technological frontier in several areas of electronics; during the 1940s and 1950s European firms remained at this frontier and were internationally competitive. Later in the 1960s, however, the European industry declined and lagged more and more behind the American industry. The situation did not improve during the 1970s and early 1980s. While Japan was able to catch up with the American industry, the European industry continued to lag behind the American industry. The European industry was not able to profit from a series of policies of support mounted by European governments. The following sections recount the history of the world semiconductor industry in order to frame the puzzle in a more detailed way.

1 A brief history of the world semiconductor industry

The semiconductor industry was (officially) initiated in 1947 with the discovery of the transistor at Bell. Semiconductor devices, however, had been produced for quite a long time even before this. The dynamics of the world semiconductor industry show a continuously evolving pattern of innovativeness and competitive advantage across firms and across countries.

1.1 *The period before the transistor*

Before the invention of the transistor, semiconductor materials had been used in several devices. Galena, for example, was used in cat's whiskers, a detector rectifier for wireless signals, and silicon was used in high frequency detectors (radars) during the Second World War. Beginning in the 1930s, copper oxide and selenium were also used in small current rectifiers; because these rectifiers were small and consumed little energy they eventually replaced vacuum diodes. Moreover, in the late 1930s and early 1940s, limited types of semiconductor diodes were both produced and used.

At the time of the invention of the transistor (1947), the amplification function was performed by triodes, a type of electron tube. Triodes were manufactured on a large scale and used mainly in radios. However, they were large, often unreliable and consumed large amounts of power. As a result, their use was limited.[1]

Before the invention of the transistor, semiconductor devices and electron tubes were produced within a stable and oligopolistic world electrical industry. This industry was characterized by the dominance of vertically integrated producers in the United States, Europe and Japan: RCA, Sylvania, General Electric, Raytheon, Westinghouse and Western Electric in the United States; Hitachi, Matshushita Electric, Toshiba, Mitsubishi Electric and Nippon Electric in Japan; Siemens, Telefunken and Bosch in Germany; Brown Boveri in Switzerland; Philips in the Netherlands; AEI, EE and GEC in Britain;[2] Thomson-Houston, CGE and CSF in France. These producers manufactured both electron tubes and semiconductors in-house and formed a rather stable oligopoly; early in the century, electrical producers who had entered the electron tube market deterred the entry of new firms by the use of patent restrictions. While anti-trust legislation in the United States later lifted some of these restrictions and allowed the entry of new firms into the industry, patent, market sharing and price agreements in Europe gave several large firms a stable oligopoly in the market for receiving tubes (OECD, 1968; Newfarmer, 1978).

Both before and soon after the invention of the transistor, European electrical firms had advanced technological capabilities in electrical and electronics technologies. Although the European industry, and particularly the German one, had been destroyed during the War and had to be rebuilt afterwards, it was able to remain at the technological frontier in several electrical and electronics technologies (Freeman, 1982). These advanced capabilities allowed European producers to become innovators and to remain among the world's major producers of electron tubes (Freeman, 1965). In 1954, for example, Mullard (Philips's subsidiary) controlled 59 per cent of the electron tube market in Britain, while two British firms, AEI and GEC, together controlled 17 per cent of this market (Golding, 1971 p. 181, and Kraus, 1968). During this period, these European producers were also very active in research for new types of electronics components. At the moment the transistor was invented in the United States, for example, French and British firms were doing advanced research on solid-state amplifiers, and were rapidly able to understand the scientific and technological importance of that discovery.

Similarly, during the 1930s and 1940s, these European firms were among the first producers of new semiconductor devices. During the late 1930s and early 1940s, for example, Siemens, GEC and British Thomson-Houston (later AEI) were among the first producers of germanium diodes; they began production at approximately the same time as General Electric began the commercial production of its germanium point-contact diodes (1938) and its silicon-point contact diodes (1941).[3] During the late 1940s, Philips, Siemens, GEC and British Thomson-Houston were also able to produce the transistor, following its discovery by the Bell Laboratories (OECD, 1968).

The above discussion clearly indicates that European producers of electronics components were not at either a technological or a productive disadvantage relative to American producers at the moment the transistor was invented; this situation contrasts sharply with that of the 1960s and 1970s.

1.2 The semiconductor industry from the early 1950s to the early 1980s

The invention of the transistor at Bell Laboratories (USA) in 1947 officially established the semiconductor industry. In the United States, the major producers in this new and rapidly growing industry consisted both of large, established, vertically integrated electron tube producers and new specialized producers. In Europe and Japan, on the other hand, only large, established, vertically integrated electron tube producers were present. In the United States, moreover, the government gave the new semiconductor industry consistent and immediate support, while in Europe and Japan the government did not offer such support.

During the 1950s American firms were the technological leaders in the world industry, maintaining a highly innovative record. Bell Laboratories, Western Electric, General Electric, RCA and Westinghouse, together with Texas Instruments, were the major innovators and the main producers.

American technological leadership in semiconductor technology at this time resulted in a major new innovation which radically transformed the existing semiconductor industry: the integrated circuit. The integrated circuit began to be produced on a large scale by American firms in the early 1960s. As a result of this innovation, the American semiconductor industry underwent major structural changes: new specialized firms entered integrated circuit production while several established vertically integrated circuit producers had to exit from the market. This industry also became the undeniable leader in the integrated circuit market at the innovative, productive, and commercial levels. American producers not only exported their production to Europe, but also made direct foreign investments in Europe.

European firms performed less satisfactorily than American firms at the innovative level, although not at the productive and commercial levels. Europe therefore remained mainly self-sufficient in semiconductors during the 1950s. Philips and Siemens, in fact, were among the major producers of transistors, maintained large shares of the European semiconductor market, and reaped huge profits from their open-market sales of semiconductors. During the 1960s, European producers began to lag more and more behind American producers in integrated circuit technology; however, they still did not exit from the semiconductor industry.

The Japanese industry, on the other hand, fell somewhere in-beween the European and American industries. Japan was not a technological leader in integrated circuits, but it did enjoy commercial success in the production of transistors during the 1960s and it was able to protect its markets from American penetration by impeding American direct foreign investments.

The rate of technological change in the semiconductor industry did not diminish in the 1970s; major innovations such as the microprocessor and the microcomputer were introduced during this period. These innovations radically transformed the industry again, altering the boundaries between the semi-conductor industry and the other electronic sectors and changing the barriers to, and the mode of entry into, the semiconductor industry. In the United States, in fact, most of the entrants into this industry consisted of either new firms which focused their activities on specific market segments, or existing users of semiconductors which integrated their R & D and production upstream. In Europe and Japan, on the other hand, the first type of entrant was almost totally absent and the second type was rather limited in number.

During this period, American firms continued to be the technological, pro-ductive, and commercial leaders in the semiconductor industry. By the early 1980s, however, they had to share this leadership with Japanese firms which had been able to close the gap in some product areas. European producers, by con-trast, were trying to re-enter the standard market for large-scale integration semiconductor devices.

2 Why did the European industry decline and why was it then unable to catch up?

Macro factors that affect the dynamics of the semiconductor industry, as well as those of other sectors of the economy, provide some answer to the puzzle about the European semiconductor industry. These factors include scientific and technological infrastructure,[4] socio-political/institutional characteristics, labour market characteristics,[5] the overall size and sophistication of demand,[6] and the type of financial system.[7] Most of the differences relating to these macro factors across Europe, Japan and the United States, however, also existed during the 1940s and 1950s, when European firms fared rather well. Therefore, these differences alone cannot explain either the decline of the European industry during the 1960s or its unsuccessful attempt to catch up with its American and Japanese counterparts during the 1970s and early 1980s.

It is argued here that the explanation of the decline and the unsuccessful performance of the European semiconductor industry must be related to a set of additional sector-specific factors concerning supply, demand and public policy. These factors include the organization of R & D and production and the strategies of firms; the structure of the industry; the composition of demand and the type and extent of government policy. The following chapters will examine the effects of these factors on the evolution of the European semiconductor industry and will try to answer the following questions:

First, to what extent did differences in the organization of R & D and production and in the strategies of European vs. American and Japanese firms affect the innovativeness and competitiveness of these firms? How did the structure of the semiconductor industry affect the creation and diffusion of new technologies and the competitive struggle among firms?

Some semiconductor firms are diversified, vertically integrated electronics final goods producers, while others are specialized semiconductor producers. Most of these producers carry out both R & D and production; few carry out mainly production. Finally, some electronics final goods firms do R & D but do not produce semiconductors. All of these different types of firms in the semiconductor industry, however, must adapt to changing conditions, face high degrees of uncertainty, and are unable to rely solely on past experience in their technological choices. As a result, their strategies will differ. Firms may concentrate on specific technological options. They may produce a wide or a limited range of standard products. They may focus on custom or specialized products. They may be imitators or innovators. Finally, they may concentrate on major innovations, minor innovations, or incremental innovations and improvements. The variety of these firms will determine the structure of the industry. This study examines how differences in firm structures and firm strategies, as well as in industry structure, affected the innovativeness and competitiveness of firms in the semiconductor industry.

Second, did the structure of demand affect the rate and direction of technological change in Europe, the United States and Japan? If so, what were the consequences for the international competitiveness of the European semiconductor industry?

During the 1950s and 1960s, the demand for semiconductor devices in Europe as well as in Japan derived primarily from electronics consumer goods production, while in the United States it derived primarily from military and computer production. During the 1970s, on the other hand, the demand for semiconductor devices converged in Europe, the United States and Japan. This study examines how the different structures of demand, and their changes over time, affected the rate and direction of technological change.

Third, what role did European public policies play in shaping both the rate and direction of technological change and the competitiveness of the European industry in comparison with the role played by American and Japanese policies?

Government policy was very much present in the United States and very much absent in Europe and Japan at the beginning of the industry. During the 1970s, however, both European and Japanese government policies increased in both size and extent. But while Japanese public policy contributed significantly to the successful performance of the Japanese industry, European public policies were not able to improve the performance of the European industry.

3 Methodology and road map

The approach followed in the following chapters on the European industry draws on a number of specific intellectual traditions. It is a dynamic, micro, evolutionary, comparative and historical approach.

It is a dynamic approach because it emphasizes changes in technologies and industries, pays attention to rises and declines of national industries, stresses changing technological advantages or varying lags among firms and countries, and underlines successful catching-up processes and failures to adapt. Such a dynamic approach is particularly important in a high-technology industry such as the semiconductor industry, since opportunities at the technological level are so high and change so rapidly. This approach dates back to Schumpeter, according to whom a theory that is static in nature and is applicable only to stationary processes cannot explain the evolving and intrinsically unstable nature of the capitalistic process (Schumpeter, 1950, 1951). It is best exemplified by the way Schumpeter analyzed the process set in motion by the break in the circular flow of economic life (Schumpeter, 1911). This dynamic approach emphasizes both the continuities and the discontinuities in the process of technological change. Discontinuities refer to radical innovations and to major changes in technologies and industries, while continuities refer to the incremental and cumulative aspects of technological change. Both of these elements play a major role in the evolution of firms and industries, as well as in the international competitiveness of countries.

It is a micro approach because firms represent the major unit of analysis. Such a micro approach allows one to specify diversities among firms that operate in a world of continuous change and in situations of high uncertainty. This micro-level method of inquiry was proposed by Schumpeter in his methodological individualism and has more recently been stressed by authors such as Simon (1957, 1984), Cyert and March (1963) and Nelson and Winter (1982).

It is an evolutionary approach because it emphasizes the adaptability, response and search processes of firms, the interaction and selection of firms in the industry, and the type of industry evolution. This evolutionary approach relates mainly to the work of Alchian (1950) and Nelson and Winter (1982). Within this dynamic, micro and evolutionary approach, particular attention is devoted to intersectoral aspects of technological change and to the structure of the industry. The intersectoral linkages represented by demand structure play a major role in affecting the rate and direction of technological change in an input industry such as the semiconductor industry, because they create a two-way transmission mechanism to induce innovation and changes in technology in both input and user industries. The structure of the industry, represented by the composition of strategic groups of firms, on the other hand, is linked to the rate of technological change by a well-defined two-way relationship.

The following chapters also apply a comparative perspective. Although the

European industry constitutes the main focus of analysis, its evolution is continuously compared with those of the American and Japanese industries. In this way, the various factors affecting the specific evolution of the European industry can be better identified and analyzed.

Finally, the approach of this book is historical, because it emphasizes the major role played by the past experiences and established technological capabilities of firms. At a general level, both Schumpeter, Usher (1954) and, more recently, Rosenberg (1972, 1976 and 1982) have established a tradition for the historical study of innovation processes and have emphasized the fruitfulness of historical investigations in this area. Schumpeter, in particular, explained the case in favour of such historical analysis in detail. For Schumpeter, 'economic life is a unique process that goes on in historical time and in a disturbed environment' (Schumpeter, 1951, p. 308). Therefore, 'we must investigate historically the actual industrial processes that produce [the fluctuations in investments] and by doing so, revolutionize existing economic structures' (Schumpeter, 1951, p. 312). Schumpeter believed that the collection and analysis of single firm histories should be the primary input in the study of innovation and, subsequently, of economic fluctuations and instability: 'what is required is a large collection of industrial and locational monographs all drawn up according to the same plan and giving proper attention, on the one hand, to the incessant historical change in production and consumption and, on the other hand, to the quality and behavior of the leading personnel' (Schumpeter, 1951, p. 314).

The next six chapters are organized in the following way: Chapter 2 provides an introduction to the semiconductor industry. This introduction includes a brief discussion of semiconductor technology, the industry and its markets, as well as an analysis of the economics of the semiconductor industry in terms of technological change and interdependency, the types of firms in the industry, and industry dynamics.

Chapter 3 discusses the major sector-specific factors (supply, demand, and public policy) that affected the evolution of the European semiconductor industry. This discussion is introductory and lays the ground for the next three chapters.

Chapters 4, 5 and 6 represent the core of the book: they reconstruct, analyze, and discuss the evolution of the European semiconductor industry at a comparative level. The history of the European semiconductor industry is divided into three periods: the transistor period (1947/51-1959/61); the integrated circuit period (1959/61-1971); and the large-scale integration (LSI) period (1971 to early 1980s). These periods correspond to the dates of the three major innovations that occurred within this industry: the transistor in 1947, the integrated circuit in 1959-61, and the microprocessor in 1971. Each period had quite different effects on the development of technology, the organization of R & D and production, the mode of entry into the industry, the competition among firms, and the consequences of government policy toward the industry.

It should be noted that within each of these periods, the diffusion of innovations among users was a long and slow process. Old and competing products continued to exist alongside new products, and firms which produced old products continued to survive alongside new entrants. It should also be noted that within each technological regime, many incremental innovations and improvements occurred throughout the history of the semiconductor industry. These developments played an important role in the evolution of the industry and in the competition among firms.

In Chapters 4, 5 and 6 particular attention is given to features of technological change, and to supply, demand and public policy factors. The focus of the analysis in these chapters is on the strategies of, and competition among, firms, the relationship between market structure and innovation, the structure of demand, and the agencies, tools and type of public policy in the European countries. Each chapter presents a comparison of the evolution of the European industry with that of the American and Japanese industries.

Finally, Chapter 7 draws some general conclusions and discusses the present state of the European semiconductor industry.

Notes

1. Braun and MacDonald (1982).
2. In Great Britain, Ferranti only became a supplier of electron tubes for radio manufacturers in 1946. English Electric took control of Marconi, transferring all tube activities into the new subsidiary English Electric Valve between 1943 and 1953. During the same period, Pye Limited began to produce television tubes.
3. Golding (1971) and interviews.
4. See, for example, Pavitt and Walker (1976) and Freeman (1982).
5. See the discussion on labour mobility in the United States and the seniority system in Japan.
6. An attribute of the American market.
7. The basic reference on this subject in Zysman (1983). He distinguishes between capital-market based financial systems (typical of the United States and Great Britain); the credit-based, institution dominated financial systems (typical of West Germany); and the credit-based, price administered financial systems (typical of Japan, France and Italy).

2 An introduction to the semiconductor industry

Introduction

The semiconductor industry is an excellent example of a growing and innovative industry. Almost non-existent during the early 1950s, this industry reached world sales of $27 bn. in 1984 (*Electronics*, January 1985). Throughout the course of its growth, moreover, this industry experienced a continually high number of major and minor innovations and technological improvements.

The semiconductor industry is a particularly important industry in modern economies. It lies at the core of, and has revolutionized, the electronics industry. Technological change in the semiconductor industry has affected technological change in the computer, telecommunications equipment, electronics consumer goods and industrial/medical/professional equipment industries. Technological change in the semiconductor industry has also allowed the electronics industry to penetrate industry and society as a whole. In fact, mini and home computers, digital telecommunication switching equipment, private branch exchange systems, industrial robots, process control equipment, as well as many other products, have been developed as a result of advances made in semiconductor technology.

Because of its crucial technological and economic importance for the electronics industry, and for industry as a whole, the semiconductor industry has also become a strategic industry for the international competitiveness of various economies. Innovation rates, productivity growth and export performance for a large number of sectors have become dependent on the rapid adoption of microelectronics technology in both products and processes. As a result, business firms and national governments have advocated increased support for this strategic industry.

In this introductory chapter, the semiconductor industry will be analysed in both technological and economic terms. In Part 1, semiconductor technology and the characteristics of the semiconductor industry and market are briefly described. The main technological principles, the major semiconductor devices, the basic manufacturing processes, the intersectoral linkages, the main markets, and the major semiconductor firms are introduced. In Part 2, the major characteristics of the semiconductor industry are discussed. Particular attention will be given to the features of technological change, interdependencies, and the types of firms and their interaction in the industry.

1 Semiconductor technology, industry and markets

In this section, a brief introduction to semiconductor technology, industry and markets is given. The reader interested in a more detailed account can choose among a wide range of reports.[1]

1.1 Semiconductors: an electronics component

Semiconductors are active electronics components. Active electronics components switch, modulate and amplify electrical signals. They include both vacuum and gas-filled electron tubes.[2] Their characteristics differ from those of passive electronic components, such as resistors, capacitors, relays and connectors (see Figure 2.1).

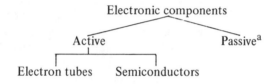

[a]Passive components consist of resistors, capacitors, connectors and relays.

Figure 2.1 Classification of electronic components

1.2 A sketch of semiconductor technology

Semiconductors derive their technological importance from the fact that they are materials which can be either conductors or insulators. At low temperatures semiconductor materials have the same basic structure as insulators. Unlike insulators, however, their conductivity varies according to changes in temperature, impurity, or optical excitation. In the last case, for example, with a process called 'doping' the addition of a controlled amount of impurities (dopants) changes a semiconductor from a poor to a good conductor of electricity. During the doping process, if a phosphorous atom (with five outer electrons) replaces the silicon atom (with four outer electrons) in the crystal, one electron remains free. This free electron makes the crystal become a negative type (*n*-type) semiconductor, able to carry electric current; the negatively charged electrons are the carriers. On the other hand, if a boron atom with three outer electrons replaces a silicon atom in the crystal, a 'hole' is created. The 'hole' makes the crystal become a positive type (*p*-type) semiconductor, able to carry electric current; the positive charges ('holes') are the carriers. When a *p*-type and an *n*-type of material are united, they form a junction. The junction is the base of the rectification, amplification and switching functions of semiconductor devices.[3]

The basic elements of most semiconductor devices are diodes and transistors.

Diodes are the simplest kind of semiconductor device; they are made up of forward and reverse biases and they have an asymmetric current flow. When two diodes are joined together a junction transistor is created. The transistor is a three terminal device which consists of the emitter, the base, and the collector.[4]

1.3 *The classification of semiconductors*

By material
There are two types of semiconductors: elemental and compound. Elemental semiconductors, such as silicon and germanium, are composed of a single species of atom which has four electrons in its outermost shell; these materials are located in Column IV of the periodic table. Compound semiconductors, such as gallium arsenide, are composed of combinations of atoms located in either Column III or V or Columns II and VI of the periodic table.

In the early periods of the industry, germanium was the most widely used semiconductor material. Germanium has a lower melting point, a lower absorption of impurities, and a higher frequency performance than silicon.

Later on, silicon became the dominant material in the semiconductor industry. Silicon is cheaper than germanium. In addition, it can operate at higher temperatures than germanium and, most important, permits the batch processing of semiconductor devices.

The third type—compound semiconductors—also have some attractive properties, such as high speed. The difficulty of obtaining exact purity balances, however, has impeded their wide use in industry.

By product
There are three major product groups: discrete devices, opto-electronic devices and integrated circuits (see Figure 2.2).

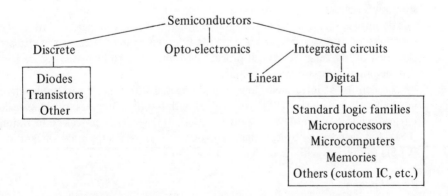

Figure 2.2 Classification of semiconductors

Discrete devices are single elements of an electrical circuit. They may be further grouped into three sub-categories: diodes, transistors and other discrete devices (such as silicon-controlled rectifiers). Diodes perform the rectification function, transistors perform the switching and amplification functions under low power conditions, and silicon controlled rectifiers perform the switching function under high power conditions.

Opto-electronic devices are used in optical systems and are usually made of gallium arsenide. The major opto-electronic device is the light-emitting diode which is used for detecting light.

Finally, integrated circuits are whole circuits placed on a silicon wafer. In this study, only monolithic integrated circuits are considered. Monolithic circuits have all the circuit elements on the same semiconductor wafer. Hybrid integrated circuits, on the other hand, have components in semiconductor technology (integrated circuits and discrete devices), other types of components and interconnections in film technology. According to function, monolithic integrated circuits can be grouped into digital or linear (analog) types. In linear integrated circuits the relationship between inputs and outputs varies continuously over a specific range. In digital integrated circuits, on the other hand, the relationship between inputs and outputs is a binary (on/off) one. The configurations of transistors which perform binary encoded operations are called 'gates'. Digital integrated circuits can be grouped into standard logic families, memories, microprocessors and microcomputers.

Memories are digital integrated circuits which can store information. They are measured in bit, or binary units of information. Types include ROM (read-only-memory), whose content cannot be changed, PROM (programable-read-only-memory), whose content can be programed after manufacture and then cannot be changed, and RAM (random-access-memory), whose content can be changed. RAM memories can be static or dynamic. Static RAMs maintain their content while dynamic RAMs must be continuously refreshed because their stored information decays over time.

Microprocessors are digital integrated circuits which can perform all of the functions of the central processing unit of a computer. Microprocessors are classified by the length of words processed (4-bit, 8-bit, 16-bit). Microprocessors and memories make up microcomputers.

The density of integrated circuits has increased with time. It can be classified in small-scale integration (SSI) for devices with less than 10 gates; medium-scale integration (MSI) for devices with less than 100 gates; large-scale integration (LSI) for devices with less than 100,000 gates; very-large-scale integration (VLSI) for devices with more than 100,000 gates.

By technology

Transistors and integrated circuits are grouped into bipolar and MOS technologies. In bipolar technology, semiconductor devices are of a bipolar type:

conduction occurs through the movement of both electrons and holes and control occurs through an electric signal. In MOS (metal-oxide semiconductor) technology, on the other hand, semiconductor devices are of a unipolar type: conduction occurs through the movement of either electrons or holes and control occurs through voltage.

Bipolar and MOS technologies have quite different characteristics. Bipolar devices are faster than MOS devices. Yet MOS devices are cheaper to manufacture, and have lower power consumption and heat generation than bipolar devices.

Bipolar and MOS devices are therefore used for different purposes. Bipolar devices are used when operating speed is required and when several analog (linear) operations are performed. MOS devices, on the other hand, are used when high circuit density and high integration is required and when digital operations are performed.

There are several families of bipolar and MOS devices. The main families are indicated in Figure 2.3. Among bipolar families, TTL (transistor-transistor logic) is a widely used 'logic' device; ECL (emitter-coupled logic) is a very high speed device; I2L (integrated injection logic) is a highly integrated device. Among MOS families, PMOS (p-channel MOS) is the oldest type of MOS device; NMOS (n-channel MOS) is widely used in memories and microprocessors; CMOS (complementary MOS) is increasingly used because of its low power consumption; SOS (silicon on sapphire) and CCD (charge-coupled device) are used in specific applications.

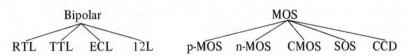

Bipolar   MOS

RTL   TTL   ECL   12L   p-MOS   n-MOS   CMOS   SOS   CCD

Source: Levin (1982), p. 15.

Figure 2.3 Classification of digital integrated circuits

By power capability
Semiconductors can also be classified according to their capability to handle power. Most semiconductor devices are designed to handle low power.

Certain semiconductor devices (the power devices), however, must have the capability to handle high voltages. They use silicon as a material, and are most often used as rectifiers (that is, as devices which conduct current in one direction only). Power semiconductors are usually used as components in electrical equipment which transform electrical power from alternating current (ac) to direct current (dc). There are several types of power semiconductors, ranging from diodes, to transistors, to thyristors. Of the various types of high-power semiconductors, two are particularly important: the silicon rectifier diode and the thyristor.

Silicon rectifier diodes are power rectifiers which can handle voltages of between 50 to 5,000 V. They are composed of two layers with one junction between the *n* and the *p* type region.

Thyristors, on the other hand, are composed of four layers. Of the various types of thyristors, the silicon controlled rectifier and the triacs are particularly important. The silicon controlled rectifier is the most widely used thyristor: it can be used both from an alternating current supply and from a direct current supply. It is used, for example, in light and heating controls. The triacs is specialized in alternating current switching power control. It is used, for example, in lamp dimmers.

By customization

Semiconductor devices can also be grouped into standard, custom and semi-custom types. Standard devices are produced in large volumes for a wide range of applications: they have a low price and a high reliability, although they do not use the space on the device in the most efficient way.[5] Custom devices, on the other hand, are produced in small lots for specific uses, following users' specifications. They have a higher price and a longer turnaround time than standard devices, but they make the most efficient use of the silicon space. Finally, semi-custom devices have some characteristics of the standard devices and others of the custom devices. They have a generic structure (like standard devices), so that they can be produced in large volume, at low price, with a high reliability and a rapid turnaround time. They can, however, be customized either by the producer or by the user, and, therefore, are able to meet the specific user's specifications and requirements. The semi-custom market has increased in the past years compared to the standard and the custom markets.

Semi-custom semiconductor devices can be grouped into gate arrays, standard cells, and silicon compilers.[6] Gate array devices leave only the final interconnection layer to be specified by the user. They have shorter turnaround and development times and lower costs than standard cells. Because of the high circuit density, they use CMOS technology. They are the most widely used semi-custom devices (see Table 2.1). They are the 'customized' solution for small volumes (less than 40,000 units) (ICE, 1983). Standard cell devices, on the other hand, are developed through the combination of specific standard cells taken from a design library, so that the overall circuit requirements and specifications can be obtained. Standard cell devices use the space of the device more efficiently, have lower production costs and are better able than gate arrays to incorporate memory and microprocessor functions into design. Even though they are used less than gate arrays, standard cells are expected to be widely used in the coming years. They represent the 'customized' solution for medium volume (between 40,000 and 100,000 units). For large volumes, full custom devices are the best solution (ICE, 1983). Finally, silicon compilers use cells in a more flexible way than do standard cells. They permit the design

Table 2.1 Application Specific Integrated Circuits (ASIC) (full custom, gate arrays, standard cells, programmed logic)

A. World Sales—1979, 1984, 1989

	World sales: open market ($ millions)	% of total integrated circuit sales
1979	490	6.9
1984	2,468	12.2
1989 (predicted)	9,053	16.3

B. Rank of the major world producers of semi-custom semiconductors in 1983

Merchant	Captive
1. Ferranti	1. IBM
2. Fujitsu	2. Fujitsu
3. Interdesign (Ferranti)	3. NEC
4. Motorola	4. Hitachi
5. Texas Instruments	5. Texas Instruments

Sources: A. Dataquest in *The Economist* (8 December 1984).
B. Gnostic Concept in *Electronics* (30 June 1983).

of more sophisticated circuits and use the space of the device more efficiently than gate arrays or standard cells. Even though they are not widely used at the present time, they will be increasingly used in the coming years.

Open market sales of application-specific integrated circuits (custom, standard cells, gate arrays and programmed logic) increased from $490m. in 1979 (6.9 per cent of total of world integrated circuit sales) to $2,468m. in 1984 (12.2 per cent) and are predicted to reach more than $9 bn. in 1989 (16.3 per cent). (Dataquest in *The Economist*, 8 December 1984, see Table 2.1). Full custom and gate arrays constitute the most sold application specific integrated circuits.

1.4 *Manufacturing process*

The manufacturing process can be divided into five stages: the mask design, the crystal growth and slice preparation, the fabrication of the device, the assembly and the final testing. Mask design is a highly time-consuming and costly process, now controlled by a computer. Crystal growth and slice preparation prepare the wafers through etching and polishing processes. Then the fabrication stage begins: it includes the growth of a single crystal layer on the substrate (epitaxy), the growth of a silicon dioxide layer (oxidation), the coating of the silicon dioxide layer with organic material (lithography), the print of the mask on the device (projection) and the introduction of impurities on the open areas of the surface (diffusion or ion implantation). Oxidation, lithography, diffusion and ion implantation may be repeated several times with different mask patterns. At the end of the repeated processes, the desired circuit structure is reproduced on the wafer. At this point, the layer is insulated and electrical connections are formed (metallization). Then metallic probes are used to discard defective parts (first test). The next manufacturing stage is the assembly, which orientates and packages the wafers, and connects them with the outside. Lastly, the final test stage includes several final electrical and mechanical tests.

Each of the five stages of the manufacturing process has different yields. Yields represent the percentage of semiconductor devices which are usable in the next stage of the manufacturing process. The wafer fabrication stage has traditionally had the lowest yield. This is shown by comparing the yields in the production of TTL integrated circuits during the late 1960s with the yields in the production of 64K dynamic RAMs in the early 1980s.

Table 2.2 Comparative yields in production of TTL and 64K dynamic RAMs

	TTL (late 1960s) %	64 KDRAMs (early 1980s) %
Wafer process yield	80	78
Probe yield	40	28
Assembly	90	92
Final testing	90	75

Sources: for TTL: Finan (1975), p. 20. For 64 KDRAMs : ICE (1982) in United Nations (1983) p. 108.

Generally, the yields of established products are higher than those of new products. In the early 1980s, the global yield of a 4K semiconductor device, for example, ranged between 35 and 45 per cent; of a 16K, between 15 and 25 per cent; and of a 64K, between 3 and 6 per cent (Sterling Hobe Corporation in United Nations, 1983, p. 107). In addition, yields vary among technologies: MOS semiconductor devices usually have higher yields than bipolar devices.

The capital investment in production facilities has increased considerably over the years. During the first half of the 1950s, the investment of General Transistor, for example, amounted to $100,000. In the late 1960s, the capital investments needed to remain successful in the semiconductor industry increased to $500,000 (Golding, 1971, p. 117). They then reached $2 million in 1972, $5 million in 1978 (Truel, 1980) and over $10 million in the early 1980s (*Journal of Electronic Engineering*, in United Nations 1983, p. 102).

Together with the cost of capital, economies of scale have also increased over the years. In the early 1980s, for example, a factory producing 16K memory devices was efficient with a production of 320,000 units per month (*Journal of Electronic Engineering*, in United Nations, 1983).

1.5 *Semiconductors and the electronics industry*

Semiconductors go into all electronics final products: consumer goods, computers, telecommunication products, industrial/instruments/medical equipment (see Figure 2.4).

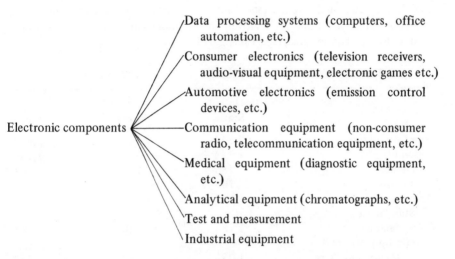

Electronic components

Data processing systems (computers, office automation, etc.)

Consumer electronics (television receivers, audio-visual equipment, electronic games etc.)

Automotive electronics (emission control devices, etc.)

Communication equipment (non-consumer radio, telecommunication equipment, etc.)

Medical equipment (diagnostic equipment, etc.)

Analytical equipment (chromatographs, etc.)

Test and measurement

Industrial equipment

Figure 2.4 The electronics industry

Electronics final products have different kinds of demands for semiconductor devices. Computers require mostly digital integrated circuits (memories, logic families, microprocessors and microcomputers) in both bipolar and MOS technologies. Consumer goods require digital integrated circuits of various types (for calculators, watches, home computers, video tape recorders, automotive equipment) and large numbers of discrete devices and bipolar linear integrated circuits (for radios, TV, and automotive).[7] Telecommunication equipment requires digital integrated circuits (mainly for switching systems, private branch exchanges, etc.) as well as discrete and linear integrated circuits (for transmission). Finally, instruments and industrial and medical equipment require discrete components and linear and digital integrated circuits.

1.6 *The markets for semiconductor devices in the United States, Europe and Japan*

In 1984 the world market for semiconductor devices amounted to $26,935m. $20,979m. of which concerned integrated circuits and $5,956m. discrete devices. The market for semiconductor devices (excluding opto-electronic devices) amounted to approximately $11,800m. in the United States, $8,200m. in Japan and $4,500m. in Western Europe.[8] The American market for standard digital integrated circuits, microprocessors, microcomputers and memories was the largest market for these products. Differences in market sizes between Europe, Japan and the United States were much smaller for discrete devices and linear integrated circuits (see Table 2.3). As a share of electronics components, semiconductors are relatively more important in the United States than in Europe and in Japan (see Table 2.4).

Table 2.3 Major markets for semiconductors, 1982 ($ millions)

	United States	Western Europe	Japan
Semiconductors	7,525	2,436	3,832
Discrete	1,322	771	1,160
Diodes	486	313	361
Transistors	709	350	664
Integrated circuits	5,942	1,519	2,372
Standard logic families	1,183	610	395
Microprocessors & Microcomputers	1,053	174	370
Memories	2,113	308	446
Linear integrated circuits	868	427	696
Opto-electronic devices	261	103	227

Source: *Electronics* 13 January 1983.

1.7 *Major end user markets and electronics final goods markets*

In 1979, the major end user markets for semiconductors were the consumer and computer markets as Table 2.5 shows.

The importance of the various end-user markets for semiconductors derives from the amount of semiconductors required in the production of various electronics final goods. In 1982, the markets for the various electronics final goods totalled approximately $147 billion in the United States, $61 billion in Western Europe and $49 billion in Japan.[9] As Table 2.8 shows, the United States constituted the largest market for most types of electronics products (especially data processing). Europe, on the other hand, constituted the largest market for communication equipment, especially telecommunication equipment.

Table 2.4 Markets for electronic components, 1982 ($ millions)

	United States	Western Europe	Japan
Semiconductors	7,525	2,436	3,832
Other components	12,792	9,425	11,937
Passive		5,174	6,647
Tubes	1,618	1,484	918
TV picture	947	1,042	675

Source: Electronics (13 January 1983).

Table 2.5 1979 World end-user markets

	Percentage of total market
Consumer-automotive	31.4
Computer	21.3
Industrial	15.6
Communications	9.6
Federal	7.2
Distributor	14.9

Source: Motorola, in the United Nations, 1983, p. 28.

1.8 *The major semiconductor firms in the world*

In the early 1980s five of the ten major corporations were American, four were Japanese and only one was European, as Table 2.6 shows. In the early days of the industry, however, European firms were much more successful than they were in 1984. In order to understand the reasons for this decline, it is first necessary to highlight the major features of the economics of the semiconductor industry. This is done in Part 2 of this chapter.

2 Technological change, intersectoral linkages and industrial dynamics

2.1 *Technological change and the semiconductor industry*

The evolution of the semiconductor industry represents one of the best examples of Schumpeterian dynamics. Schumpeter interpreted the nature of capitalism as a discontinuous and evolving process that is in a constant state of disequilibrium. 'Capitalism . . . is by nature a form or method of economic change

Table 2.6 Major semiconductor producers

	1982[a] Share of world production %	1984[b] Production ($ millions)
Motorola (USA)	7.4	2,255
Texas Instruments (USA)	6.9	2,350
NEC (Japan)	6.2	1,985
Hitachi (Japan)	4.5	1,690
National Semiconductor (USA)	3.9	1,270
Toshiba (Japan)	3.8	1,460
Intel (USA)	3.5	1,170
Philips (Europe)	2.8	1,150
Fujitsu (Japan)	2.5	815
Fairchild (USA)	2.3	na

Source: [a]ICE, in United Nations, 1983, p. 131. [b]ICE in Electronics Week (1 January 1985).

and not only never is, but never can be, stationary' (Schumpeter, 1950, p. 82). Capitalism's 'evolution is lop-sided, discontinuous, and disharmonious by nature . . . . Disharmony is inherent in the very *modus operandi* of the factors of progress' (Schumpeter, 1939, p. 103). For Schumpeter, therefore, disequilibrium is not the exception in the economic process: it is the rule.

In a Schumpeterian fashion the semiconductor industry has experienced continuous change due to a long string of innovations which have punctuated its entire evolution. According to Schumpeter, innovations consist of carrying out 'new combinations of productive means' (Schumpeter, 1934, p. 66); that is, introducing new goods and new methods of production, opening new markets, conquering new sources of supply and creating new organizations within industry (Schumpeter, 1950).

Several of the innovations that occurred in the semiconductor industry are typical Schumpeterian innovations—i.e., major innovations that disrupted and altered the economic system more drastically than did incremental innovations.[10] In the semiconductor industry, the invention of the transistor revolutionized the existing electronics industry and represented a major shift away from the use of amplifiers based on receiving tubes. Later on, the introduction of the integrated circuit drastically altered the existing semiconductor technology based on transistors. Finally, the introduction of the microprocessor opened completely new fields of applications for semiconductor devices.

In addition, the high number of innovations which occurred in the semiconductor industry represents what Schumpeter identified as the clustering of innovations in the industrial system. At a general level, according to Schumpeter, 'innovations do not remain isolated events, and are not evenly distributed in time, but . . . on the contrary they tend to cluster, to come about in bunches . . . Innovations are not at any time distributed over the whole economic system at random, but they tend to concentrate in certain sectors and their surroundings' (Schumpeter, 1939, pp. 100-1). For Schumpeter, the economic waves of the last centuries are linked to innovations clustered in specific sectors: textiles first, iron and steam later, electricity and chemicals still later (Schumpeter, 1951, p. 63). Therefore, Schumpeter emphasizes that innovations are historic phenomena which occur irregularly in time and across sectors.[11] In the semiconductor industry, the clustering of innovations has originated and shaped the modern electronics industry. The transistor, the integrated circuit and the microprocessor have constituted the basic components of most electronics final products (computers, telecommunications, industrial equipment, consumer products). As a result, innovations in semiconductor devices have had strong repercussions for innovations in electronics final products. In turn, because these electronics final products have become important inputs in most sectors of the economy, innovations in semiconductor devices have helped to shape industrial change throughout the economic system.

It is misleading, however, to claim that the evolution of the semiconductor industry has only been characterized by major innovations; incremental innovations and production engineering improvements have also played a major role in shaping the rate and direction of technological change. At a general level, the early Schumpeterian tradition disregarded incremental innovations and production engineering improvements as major factors. This tradition inherited Schumpeter's interest in major innovations, without noticing that, in some passages, Schumpeter also recognized the importance of minor innovations.[12] Only recently, in fact, has the importance of incremental innovations in shaping the rate and direction of technological change been recognized.[13] However, while major innovations are phenomena discrete in time, incremental innovations are phenomena continuous over time. As a consequence, incremental innovations are more difficult to identify and analyse than are major innovations. In the semiconductor industry, therefore, while major product and process innovations have been identified and numbered by Tilton (1971), Wilson, Ashton and Egan (1980) and others,[14] incremental innovations and product engineering improvements over these major innovations have not. These types of innovation, however, more than the major innovations, have been the source of the technological advantages and international competitiveness enjoyed by firms.

Following this broad perspective on innovation, the diffusion of technologies in the semiconductor industry should be regarded as a process during which a new technology replaces an old one but,[15] in doing so, encounters resistances and goes through change and improvement. The history of the semiconductor industry is rich with examples. Two are related to the period after the introduction of the point-contact germanium transistor and of the planar integrated circuit. The beginning of the production of a semiconductor amplifier, such as the point-contact germanium transistor by existing electronics firms was delayed or opposed in many cases because of vested interests in receiving tubes departments of these firms. Once the point-contact transistor was introduced (Western Electric, 1951), however, its diffusion among semiconductor producers led not only to improvements in terms of performance, reliability and cost, but also to the introduction of the grown junction transistor (Western Electric, 1951), the alloy junction transistor (General Electric and RCA, 1952) and the surface barrier transistor (Philco, 1954), (Tilton, 1971). Similarly, the beginning of the production of silicon integrated circuits was delayed or opposed by several firms that were successful in the production of transistors. Once silicon integrated circuits were introduced (Fairchild and Texas Instruments, 1961), however, their diffusion led to major improvements over the original device in terms of performance and cost, and to the development of a whole series of new integrated circuit families (see, for example, the lists in Wilson, Ashton and Egan, 1980, and Braun and MacDonald, 1982).

Technological advance in the semiconductor industry, therefore, has resulted in a continuous change of products and processes, rather than in a few major innovations introduced by a limited number of firms and replicated mechanically by the rest of the industry. This is demonstrated by the short life span of most semiconductor devices:[16] even before most types of semiconductor device entered the phase of maturity and decline, in fact, new types of products or processes had been developed to replace them.

During the course of the last thirty years, technological advance in the semiconductor industry has been characterized by three major technological regimes: the transistor, the integrated circuit, and the large-scale integration (LSI) microprocessor. Technological regimes concern 'technicians' beliefs about what is feasible, or at least worth attempting' (Nelson and Winter,1982, p. 258). Technological regimes guide the search and R&D processes of firms. They delimit technological advances in well-defined areas, because in those areas scientists, technicians and engineers have a better (or at least some) knowledge of the properties of materials, of the workings of specific technologies and of the characteristics of would-be products and processes. The three major technological regimes in the semiconductor industry correspond to the three major innovations that occurred during the history of the industry. Each new technological regime represented a major change with respect to the preceding regime, because it altered the previous area of research. As a result, each new technological regime had a profound impact on the evolution of the semiconductor industry.[17] In the following chapter, the history of the semiconductor industry will be divided into three periods, according to these technological regimes.

Technological regimes have delimited technological trajectories in the semiconductor industry. At a general level, technological trajectories relate to the advances of technologies in well-defined directions (Nelson and Winter, 1982). Technological trajectories are the result of local and cumulative advances: that is, current innovations are based on innovations that directly preceded them. They build upon, modify (more or less drastically) or simply improve existing products or processes, but, in so doing, follow some criteria or have some specific focus. In the semiconductor industry during the transistor period (the 1950s), the path was directed toward reliability and performance; during the integrated circuit period (the 1960s) toward miniaturization; and during the large-scale integration period (the 1970s and early 1980s) toward miniaturization and integration.[18]

The cumulative nature of technological advances in the semiconductor industry is evident in every period. During the 1950s and early 1960s, for example, the planar process (1960) built upon the oxide-masking and diffusion process (1955); and the epitaxial process (1960) built upon the planar process. During the 1960s, for example, bipolar integrated circuits were first resistor-transistor logic (RTL) types (1961); later they were modified to diode-transistor logic (DTL) types (1962); and subsequently to transistor-transistor logic (TTL) types (1964). In more recent times, for example, MOS Dynamic Random Access Memories (DRAM) were of 1,000 bits (1K) in 1970, 4K in 1973, 16K in 1976, 64K in 1978, 256K in 1983–4. Microprocessors were 4-bit in 1971, 16-bit in 1973, and 32-bit in 1980.

Undoubtedly, technological advance in the semiconductor industry has been influenced by high opportunity conditions. At a general level, opportunity conditions relate to the supply side of the innovation process, and have been emphasized by the 'technology push' approach: the stock of scientific and technological knowledge available to firms is responsible for generating innovations. In the semiconductor industry, opportunity conditions were favourable at the beginning of the industry, because of the advances that had been made in solid-state physics: these advances made the invention of the germanium transistor (1947) and the introduction of the silicon transistor (1954) possible (Mowery, 1982). The advances in solid-state physics, however, could not generate these innovations alone. The technical success in the innovations also required successful solutions to several practical or unexpected problems faced by innovators during the development of the new devices.[19]

With the development of the semiconductor industry, however, and particularly with the introduction of the integrated circuit, opportunity conditions relied less and less on basic scientific knowledge and more and more on technological and engineering capabilities. The integrated circuit, the microprocessor, and the memory devices, in fact, did not result from new scientific knowledge, but rather were the outcome of technological and engineering efforts of various semiconductor producers.

At the firm level, technological change in the semiconductor industry has been the result of both the R&D policy of firms and learning-by-doing and by-using processes. The R&D policy of firms has resulted in many major and minor innovations, while learning-by-doing and learning-by-using have resulted in more incremental and local types of technical advance.

Semiconductor firms devote considerable resources to R&D. During the period 1976-81, for example, the American semiconductor industry had a R&D/ sales ratio that ranged from a minimum of 6.3 per cent (1976) to a maximum of 8.5 per cent (1980) for semiconductors and from a minimum of 6.7 per cent (1976) to a maximum of 8.7 per cent (1981) for integrated circuits.[20] In addition, the threshold of R&D expenditures for a competitive production of semiconductors has increased considerably in absolute terms during the 1960s and the 1970s. Just in the period 1962-8, the threshold R&D in integrated circuits increased from $275,000 to $2 million.[21]

As previously mentioned, however, a relevant part of technological advance in the industry has been the result of learning-by-doing. Learning-by-doing takes place during the course of the R&D and production process: it refers to improvements in the organization and the performance of the production process, and in the design and characteristics of the products. It is the result of the cumulative experience of engineers, technicians, skilled personnel and labour at the R&D and production level. It refers, therefore, to incremental and local technical change, within well-defined technologies.[22]

In the semiconductor industry, learning-by-doing has generated learning curves for single products and for groups of products. The first type of learning curve is closest to the concept of learning-by-doing, because it refers to increases in yields at various manufacturing stages for each single product. For example, the average selling price of a 64K RAM in the United States was $30 after 190,000 units had been sold by American firms at the end of 1980, and declined to $7 after 3,806,000 units had been sold by American firms at the end of 1981 (SIA in United Nations, 1983, p. 113).[23] The second type of learning curve is not only the result of improved yields in specific products, but also of incremental innovations and increased miniaturization and integration. This last trend is described by the 'Moore Law', according to which the number of components on a single semiconductor device doubles every year.[24] Some studies have estimated the slope of the learning curve at the aggregate level for major product groups. Webbink (1977), for example, has found that from the early 1950s to the mid-1970s the reduction in unit cost was very similar for silicon transistors, digital and linear integrated circuits (with regression coefficients between -0.40 and -0.50) and was greater than that for germanium transistors (with regression coefficients of -0.25). The learning curve concept has played a major role in the competition among semiconductor producers, because it has given an advantage to early producers. It must be remembered, however, that the competitive advantages of firms that stem from the learning curve are nullified if experience cannot be appropriated and if completely new products or processes are introduced.

2.2 *Intersectoral linkages and interdependencies*

Technological change in the semiconductor industry can be better understood if the semiconductor industry is regarded as an input supplier with linkages and interdependencies to those sectors that use semiconductors. (The relationship between the semiconductor industry and the suppliers of semiconductor materials and production equipment is not analysed here. On this subject, see, for example, United Nations (1983).) Directly, semiconductors have vertical linkages with several downstream sectors because they are used in all types of electronics final goods and equipment. Indirectly, semiconductors have horizontal linkages with an increasing number of sectors, because of the ever expanding use of electronics final goods and equipment within the economic system.

The linkages of the semiconductor industry with other sectors create an inter-sectoral transmission mechanism for technological change. Technological change is transmitted to the other sectors as new semiconductor devices are used in electronics final goods and equipment. Semiconductor devices, for example, go into consumer products, computers, telecommunication equipment and industrial/medical/professional equipment.

The semiconductor content of electronics goods has increased from 5.8 per cent (1970) to 7.5 per cent (1980) (United Nations, 1983, p. 27). However, every electronics final goods sector affects the semiconductor industry differently, both in terms of amount and in terms of the type of semiconductor required. For example, in 1975, for every dollar of equipment produced, the value of semiconductors activated by this production was respectively:

Table 2.7 Semiconductor content (in value), 1975

Computers	4.0%
Consumer-auto.	3.8%
Telecommunications	2.5%
Government-military	2.0%
Others	1.3%

Source: Gnostic Concept in FAST (1980) p. 43.

Of course, the semiconductor content of products also varies greatly within each of these sectors. In 1973–6, for example, within the computer sector, semiconductor content was between 3 and 4 per cent for mainframe computers and between 15 and 20 per cent for minicomputers; within the consumer sector, it was between 8 and 10 per cent for colour TVs and between 50 and 70 per cent for pocket calculators.

The intersectoral transmission of technological change, however, is more broad, more subtle, and deeper than the simple mechanical input–output linkages between semiconductors and downstream users. As previously mentioned, the diffusion of new devices is neither an automatic nor an immediate process, because it involves both learning and the replacement of what is old and known with what is new and unknown. In addition, within each input–output linkage, there are interdependencies created by bottlenecks, complementarities and triggering effects which accelerate or impede the transmission mechanism.[25]

At a general level, the intersectoral transmission of technological change in a broader sense includes impulses for innovations which go in two ways: upstream and downstream. Even though they can be associated with the backward and forward linkages of Hirschman (1958) and the technology flow matrix of Scherer (1982), these upstream and downstream impulses to innovation include a broader range of interdependencies.

At a general level, upstream impulses for innovations originate from interdependencies which are deeper than the mere size of the input–output backward linkages discussed above and of the value of R&D activated in the upstream industry by an increase in the demand in the downstream industry.[26] Upstream impulses also include the pull on innovation generated by the change and past growth of demand (the so-called 'demand pull' approach) and by the expectations of firms about future trends and the growth of demand. In several instances, in fact, not only past growth, but also the expected growth of demand, focuses the innovative activity of firms in certain directions.[27]

In the semiconductor industry upstream technological interdependencies played a major role in pulling innovation in several cases. The R&D effort which led to the invention of the transistor was stimulated by the belief of the management of Bell Laboratories in the possibilities of substituting mechanical relays with electronic switches based on solid-state physics in the growing market for telecommunication equipment. The introduction of the planar transistor and integrated circuit aimed to satisfy the (potential) American military demand for miniaturized semiconductor devices for aircraft, missiles and detection systems (Levin, 1982; Mowery, 1983).[28]

Finally, upstream impulses for innovation in the semiconductor industry have originated from learning-by-using. Learning-by-using takes place, not during the manufacturing process, but during the utilization of product by the user.[29] It is particularly important when reliability, performance and system complexity are characteristic of the product. In the semiconductor industry, learning-by-using played an important role in components like power devices and certain types of LSI devices: the former must have a high reliability and performance while the latter must be integrated into complex computer systems.

What is true for upstream impulses is also true for downstream impulses. At a general level, downstream impulses for innovation do not derive only from the forward input–output linkages discussed before or from the R&D embodied in the products purchased by end-users. Downstream impulses may open up new technological opportunities in downstream industries.[30] These new opportunities then generate a new string of innovations which develop independently of the technological developments in the upstream industry.

In the semiconductor industry, intersectoral R&D flows and downstream impulses for innovation have played a major role. In the first case, and for the electronics component sector as a whole, Scherer (1982) has calculated that in 1976 in the United States, of the $594.9m. of company-financed R&D outlays, $386m. were embodied in electronics component technology; $43.7m. were embodied in products purchased by telecommunication producers; and $50m. by defence and space (Scherer, 1982). In addition, in the semiconductor industry, several innovations set in motion a series of other innovations in electronics final goods and equipment and eventually created new markets for new products. The transistor, the integrated circuit and the microprocessor, for example, created new generations of computers. Each new generation experienced a subsequent high rate of innovation in design, architecture, computational capabilities and other computer characteristics, largely independent of advances in semiconductor devices. LSI devices led to the mini and the home computer, which were addressed to a quite different market (the consumer market) from the mainframe computer market. LSI devices, in addition, made the introduction of electronic switching systems in telecommunications possible. This new product, then, originated a string of innovations in telecommunication equipment. More recently, LSI devices made the convergence of the telecommunications, computer, and consumer sectors possible, and originated a new string of product innovations aimed at new markets such as the office, the family and small businesses, which are quite different from the traditional telecommunication and computer markets.[31]

At a general level, because of interdependencies among sectors in the innovation process the locus of innovation does not necessarily lie where the actual production of the innovative product takes place. As Von Hippel (1982) has emphasized, the functional locus of innovation depends on the level of appropriability of innovations by the manufacturers or the users and on the quantity of products supplied by the manufacturers and used by each user. If the users are able to appropriate the benefits from an innovation and use few, highly specialized products, they then become the source of innovation. This is particularly true for innovations in machinery and instrumentation (Von Hippel, 1976, 1978). In the semiconductor industry, for example, Von Hippel (1977) discovered that a large number of process innovations in the production of semiconductors were developed or dominated by users.

But even when the locus of innovation lies in the semiconductor industry, the control of the innovation may not. This occurs when innovations in the semiconductor industry are so important for innovations in downstream products that downstream users decide to control the innovative process in semiconductors.[32] This took place, for example, in the production of some types of custom integrated circuits. In several cases, the decision to control the innovative process in semiconductors by downstream users led to the vertical integration of the R&D and production of semiconductors and electronics final goods. In this case, a single downstream corporation not only controlled, but also internalized the innovative process in semiconductors.

2.3 Innovation and the semiconductor firm

Firms are the major actors in the process of technological change in the semiconductor industry. During the history of the industry, firms have developed and introduced most innovations. Government laboratories and academic institutions have performed mainly basic research, but they have until recent times contributed only slightly to the direct introduction of innovations.

Throughout the history of the semiconductor industry, firms have had to face a situation of continuous change, uncertainty and (technological) complexity. While change has been a constant characteristic of the industry since its beginnings, uncertainty was radical in the early years and subsequently declined, while (technological) complexity has gradually increased. The effects of these three factors on the organization, structure and behaviour of firms are discussed in the following pages.

In the semiconductor industry change has always been the rule, not the exception. Change has been generated both endogenously and exogenously. In the first case, it occurred through continuous product and process innovations and through the competitive dynamics of the firms in the industry. In the second case it was induced mainly from changes in public policy towards the industry.

In the semiconductor industry, uncertainty has been closely associated with change. Uncertainty in the semiconductor industry has had a number of different origins: general macroeconomic conditions, industry dynamics and the outcome of firms-specific R&D projects. In this last case, uncertainty could be of a technical or a market type.[33] In addition, the level of uncertainty associated with R&D and the technological efforts of semiconductor firms ranged from very high in the case of radical product and process innovations to very low in the case of product or process improvements or in the case of the adoption of established process innovations. An example of a high level of uncertainty is provided by the invention of the point-contact transistor; an example of a low level of uncertainty is provided by the introduction of a new type of linear integrated circuit which increases the frequency range by the adoption of the latest type of photolithographic equipment.

Finally, complexity has been an increasing feature of semiconductor technology. While complexity remained limited during the 1950s, because semiconductor devices were of a discrete type and, therefore, their architecture and technological characteristics were simple, since the mid-1960s complexity has gradually increased. This increase has been the result of a technological trajectory towards miniaturization and integration. By the early 1970s, in fact, LSI devices had very sophisticated production processes and architecture, and had become electronic systems themselves. Semiconductor producers had to be able to master not only hardware but also software and system know-how.

At a general level, change, uncertainty and (technological) complexity have several consequences on the organizational structure and behaviour of firms.

These factors affect the boundaries of firms because they increase the level of transaction costs. Firms are organizations that may encompass more than one function (production, R&D, marketing, or all three functions) or more than one sector (firms may be specialized, vertically integrated or diversified). Firms can trade their products in external competitive markets. This trade, however. may be difficult to accomplish in the presence of transaction costs. The exact level of these costs, however, is determined by the level of uncertainty and the number of firms in the market, on the one hand, and by the capability of humans and organizations to formulate solutions to problems and the opportunistic behaviour of humans on the other.[34] When these factors characterize the exchange of a product between two technologically related stages of production, firms will prefer to integrate vertically (organization within the firm) rather than to use contingent claim contracts (organization within the market) in the exchange of products[35] (Williamson, 1975).

At a general level, by integrating vertically, firms try to avoid situations in which (following the formation of a contract) a small number of firms are bound together as a result of investments in R&D and production facilities and in which information is asymmetrically distributed because one firm has more information about an exchange than another. In these situations contingent claim contracts between firms are too costly and incomplete, sequential spot contracts are subject to opportunistic behaviour and inside contracts are exposed to opportunism and information impactedness. Therefore, as long as input trans-actions are not standardized and transactions are recurrent, firms have an incentive to integrate vertically (Williamson, 1975).[36]

During the history of the semiconductor industry the high level of transaction costs played an important role in inducing firms to remain or to become vertically integrated.[37] It represented a rationale for the vertical integration of the R&D stage with the production stage and of the integration of the R&D and pro-duction in one sector with the R&D and production in another, technologically related sector. The first case has been very common in the semiconductor industry. Almost all semiconductor producers have performed these two stages in-house (only recently have design houses and silicon foundries begun to emerge). The second case has characterized firms producing semiconductors for use in their electronics final products. Firms like IBM, NEC and Philips, for example, have been involved in both R&D and production of either part or all of the semiconductor components which they use in their final goods.

At a general level, changing industry conditions imply that firms must be adaptable in order to survive. Schumpeter discussed the concept of adaptation by considering the types of responses (adaptive or creative) that firms can have in the face of changing conditions (Schumpeter, 1951). He emphasized that failures to adapt are a major cause of the decline of firms: 'No firm which is merely run on established lines, however conscientious the management of its routine busi-ness may be, remains in capitalistic society a source of profit.' (Schumpeter, 1939,

p. 95). In Schumpeter's view the adaptive response is a mechanical response to change. More recently, this Schumpeterian view of adaptation has been enlarged in order to encompass the innovative response as well. Alchian, for example, has interpreted the economic system as an adaptive mechanism and has stressed that adaptation is the key for the survival of firms (Alchian, 1950). Other authors have studied adaptation at the theoretical level (Day and Groves, 1975; Nelson and Winter, 1982) and at the organizational level (Burns and Stalker, 1961).

The history of the semiconductor industry is full of examples of firms that prospered because they were adaptable or that declined because they were not adaptable. In the American industry, for example, firms such as Texas Instruments and Motorola have been able to adapt and to switch rapidly to new technologies during the course of the last twenty years, while firms such as Transitron did not abandon the old technology (gold-bonded diode) for the new one and failed.[38]

At a general level, uncertainty and change have also implied that firms in the semiconductor industry have a subjective type of rationality. In a situation of uncertainty and change, firms cannot have a perfect knowledge of either the objective random distributions of technological opportunities in the industry, or of the change of these distributions over time. This is a situation of true uncertainty in the sense used by Knight (1921), compared to one of insurable risk; while in the latter case firms can maximize expected values from objective random distributions, in the former situation firms base their decisions on subjective probabilities. Subjective probabilities indicate probability distributions which are highly subjective and based on few, if any, observations.[39]

When (technological) complexity is added to uncertainty, the rationality of firms also becomes limited (bounded). While subjective rationality is a consequence of the conditions surrounding firms, limited (bounded) rationality refers to the information-processing capability of firms. Because of either insufficient capability or high cost, firms do not calculate all of the possible alternatives, but tend to concentrate on only a few of them.[40]

The subjective and limited (bounded) rationality of firms implies that firms may make a wrong choice of technologies. Firms, in fact, may make different subjective judgements and evaluations about the future developments of various technological options. They base their decisions on their observations concerning the probability distribution of technological opportunities, their understanding of the characteristics of present-day technology, their technological capability, their success, and their managers' and technicians' technological training and attitudes.[41]

The semiconductor industry is rich with examples of technological choices which proved wrong. During the early days of the industry, germanium continued to be preferred over silicon by several semiconductor producers, even after silicon semiconductor devices were in widespread use. In the early 1960s the integrated circuit was judged wrongly by several firms that believed it would have only limited use. In the late 1960s custom LSI devices were believed to constitute the bulk of future digital integrated circuit demand. Finally, bubble memories were considered widely applicable in the computer industry.[42]

The subjective and limited (bounded) rationality of firms as organizations in a changing environment takes the form of routines. Routines are 'repetitive patterns of activity in an entire organization; [they] are a persistent feature of the organism (firm) and they determine its possible behavior.' (Nelson and Winter, 1982, pp. 14, 97). Routines include heuristics such as firm strategies and innovation processes. They perform several functions, including organizational memory, control, replication and imitation (Nelson and Winter, 1982). Moreover, because firms are open systems of coalitions, routines may be the result of truces among groups within firms (Cyert and March, 1963). They represent the conservation and the continuity of an organization such as the firm (Nelson and Winter, 1982).[43]

Among routines, strategies play an important role in shaping firm behaviour. Strategies represent the long-term plans of firms for profitability and survival (Caves, 1980). Because they are based on the subjective and limited (bounded) rationality of firms, they are greatly affected by managerial attitudes. Because they are the product of an organization, they are also influenced by firm structure. Therefore, because managerial attitudes and organizational structure varies across firms, so will strategies.

Firms will therefore differ greatly in their routines and strategies. Some firms will take more risk, while others will be more conservative.[44] Some firms will be innovators, while others will be imitators.[45] Some firms will concentrate on standard products, while others will produce custom products. All firms, however, evolve and may modify their routines and strategies in the course of their histories.

However, it is also useful for the purpose of the present analysis to group firms according to strategic groups. Firms within a strategic group adopt similar strategies, react similarly to problems and recognize mutual dependence (Porter, 1979; Caves and Porter, 1977). Some studies have used this approach in the analysis of the American and European semiconductor industries. Wilson, Ashton and Egan (1980) have grouped American semiconductor producers into four groups according to the importance of innovation and to the breadth of the product line. They have found that firms such as Texas Instruments, which aim at major innovations and have a full line of products, perform better than do firms such as Signetics, which aim at minor innovations, or firms such as Intersil which are specialized in a few product lines. The explanation for the better performance of firms which follow major innovation strategies lies in the fact that they take more risk, but also have higher profits when successful than do firms which follow different strategies. When the former also have a full line of products, they can balance their overall performance through the varying performances of the entire line of products. Sciberras (1977), in his study of the British market, grouped firms into two groups: those producing for the standard market and those producing for the custom market. He found that firms which focus on the standard market have greater overall market shares and total profits and a higher rate of innovation than do producers who focus on the custom market, although the latter group may have higher profit margins.[46]

In addition, in the semiconductor industry the concept of strategic group is implicit in the analysis of the role of semi-custom producers and silicon foundries. The strategic group of semi-custom producers is different from the strategic group of standard and custom producers. This semi-custom group is able to supply standard products that can be changed into custom products by adaptation to the specific needs of customers. The strategic group of silicon foundries, on the other hand, is different from the strategic group of design houses and firms doing both the R&D and production of semiconductors. This group is able to produce several small lots of custom LSI and VLSI semiconductor devices by using masks and designs provided by the customers.

At a general level, the routines and the strategies of firms do not necessarily remain the same over a long period of time. Rather, they evolve. Firms, in fact, evaluate the performance of their routines in terms of specific targets such as profits. If they reach their targets firms will not change their routines (Cyert and March, 1963; Nelson and Winter, 1982). It should be noted, however, that a lack of change in routines does not necessarily mean that there is also a lack of change in the products or in the techniques of firms. Routines may represent different types of innovation policies, such as policies based on incremental innovations in certain product ranges or policies focused on the imitation of the latest technology. Therefore, it is necessary to draw a distinction between firms that do not change routines and remain innovative but that change routines in order to try to survive.

When firms do not reach their targets they search for new routines that may perform better. The process of search is irreversible, contingent and local. It is irreversible because the cost of holding information is lower than the cost of acquiring it. It is contingent because it is influenced by institutional and historic factors. It is local because R&D projects are limited in number and are localized near the techniques already used by the firm.[47]

The history of the semiconductor industry abounds with examples of unsuccessful semiconductor producers which engaged in search. During the 1950s, for example, the American firm Philco engaged in search once its surface barrier transistor proved unsuccessful. The same occurred with Fairchild and with several European corporations in more recent times when their production of bipolar integrated circuits turned out to be unsuccessful.

At a general level, in industries with a high rate of technological change firms develop and accumulate firm specific knowledge and capability through subjective rationality, routinized behaviour, local search and cumulative improvements. This is related to the fact that firms are characterized not only by physical assets and manpower, but also and particularly, by know-how, intangibles, and indivisibilities.[48] Through their activity and experience, firms learn by doing, by using, and by searching. They also appropriate and accumulate know-how.

This accumulated knowledge, however, is highly firm-specific due to the different histories and experience of firms.

In the semiconductor industry, firms within each strategic group have been able to retain their advantages in certain technologies over long periods of time because of firm-specific technological knowledge and firm-specific capability. Intel, for example, has retained its advantages in MOS memories and microprocessors; Fairchild in bipolar memories; Motorola in discrete devices; Toshiba in CMOS integrated circuits; Siemens in power semiconductor devices; SGS-ATES in power linear integrated circuits.

2.4 *Industry dynamics: an introduction*

The overall dynamics of the semiconductor industry can be interpreted as evolutionary with strategic groups and firms undergoing a selection process.[49] The selection process implies that strategic groups and firms with routines that are better suited to the competitive environment survive and grow while others diminish in size and eventually disappear. Within each country, this selection process may have different degrees of intensity, as it can be pushed or hampered by several factors.

The overall dynamics of the semiconductor industry can be affected by the type of public policy, by the level of barriers to entry, by the opportunity conditions, by the appropriability conditions, by the level of demand, and by the structure of the industry. These factors affect the competitive mechanisms in the industry as well as the rate and direction of technological change.

At a general level, different types of public policy and different heights of barriers to entry may either facilitate or hamper competitive dynamics in the industry. Low barriers to entry and a type of public policy which fosters competition among firms may increase the competitive dynamics of the industry more than will high barriers to entry and a type of public policy which protects a limited number of established firms.

In addition, high opportunity conditions lead to higher rates of innovation and higher competitive dynamics in the industry than do low opportunity conditions. In the semiconductor industry, for example, opportunity conditions remained high throughout the history of the semiconductor industry because of the advances that were made in solid-state physics and because of the continuous improvements that were made in semiconductor technology (Rosenberg-Steinmuller, 1980; Levin, 1982). Because of these high opportunity conditions, the rate of innovativeness in the industry remained high and industry dynamics remained intense.

At a general level, some degree of appropriability of innovation induces a higher level of industry dynamics and a higher rate of technological change than do either no appropriability at all or full appropriability.[50] In the semiconductor industry, for example, appropriability remained low at the patent protection level because of the shortness of the product cycle, the ease of circumventing semi-conductor patents through 'inventing around',[51] and two special historical circumstances: Bell's liberal patent-licencing policy for the transistor and Fairchild's and Texas Instruments' claims on integrated circuit patents.[52] As a consequence, the key patents were widely available and subsequent patents did not effectively protect inventions. New technological developments, moreover, could circulate rather freely in the industry and technological diffusion and industrial dynamics were favoured. During the history of the semiconductor industry, however, the appropriability of the returns of innovations has increased at the firm-specific level because of the increasing complexity and system characteristics of semiconductor devices, and the importance of learning curves and lead times.

In addition, at a general level, a large and growing demand may induce higher industry dynamics and higher rates of technological change than may a small and stagnant demand. Throughout the history of the semiconductor industry, for example, demand has been rapidly growing,[53] first, because of the sub-stitution process of transistors for receiving tubes and later because of the expanding demand for electronics final products and because of the pervasive-ness of microelectronics throughout the economic system. The high growth of demand has induced innovation and competition among semiconductor firms because successful innovations provided innovative firms with large and growing sales and profits.

Finally, at a general level, the structure of an industry in terms of the num-ber and type of firms and the composition of strategic groups greatly affects the rate of technological change in an industry. The relationship between the structure of an industry and the rate of technological change, however, is rather complex because it is a two-way relationship. Schumpeter claimed that imper-fect competition (in the form of monopolistic competition, oligopoly or mono-poly) is better than perfect competition at producing a high rate of innovation.[54] Once innovation has occurred, it produces entrepreneurial profits and creates a specific market structure which Schumpeter often called monopolistic com-petition (Schumpeter, 1939, 1951). In more recent years this two-way relation-ship between market structure and technological change has been discussed extensively.[55] There is now a wide consensus that some intermediate level of concentration in the structure of the industry is more suitable to a high rate of technological change. A medium rate of concentration, in fact, produces more competition and rivalry among firms than does a monopolistic situation, but it also permits a higher level of appropriability and profitability than does perfect competition.

Chapters 4, 5 and 6 present an analysis of the specific evolution of the European semiconductor industry in a comparative perspective. Before begin-ning this analysis, however, it is necessary to isolate some sector-specific factors that have played a major role in the European industry. This is done in the next chapter.

Table 2.8 Markets for electronic equipment, 1982 ($ millions)

	United States	Western Europe	Japan
Data processing	52,145	18,025	12,130
Communication	7,082	10,084	2,513
Telecommunication	1,800	6,062	494
Consumer	21,375	15,887	10,926
Automotive	2,536	na	1,521
Test and measure	3,611	1,027	752
Industrial	3,837	2,758	3,528
Medical	3,444	1,453	1,049
Analytical	1,156	na	405
Power supply	903	331	410
Federal	31,050	−	−

Note: na = not available.
Source: *Electronics* (13 January 1983).

Notes

1. Streetman (1980) provides an advanced explanation of solid-state physics and semiconductor devices; *Science* (18 March 1977) and *Scientific American* (September 1977) discuss semiconductor technology and devices in a more concise and introductory way; Department of Commerce (1979), Braun and McDonald (1982), Hazewindus (1982) and US Office of Technology Assessment (1983) have specific parts or appendices within broader reports on the industry; finally Tilton (1971), Golding (1971), Finan (1975) and Levin (1982) give a concise discussion of the major technological features and the main semiconductor devices.
2. Vacuum tubes can be grouped according to functions in receiving tubes, cathode-ray tubes and special tubes. They can also be grouped according to their number of electrodes: diodes, triodes, tetrodes, and so on. The vacuum diode resulted from the finding that an electric current could flow from the cathode to the anode but not in the opposite direction; it was produced from the vacuum technique that was used in the manufacture of light bulbs.
3. If the p-region is biased with a positive external voltage relative to the negative region, the current will flow in the positive to negative direction; a forward bias and a forward current are said to be present. If the p-region is biased with a negative external voltage relative to the negative region, however, no current will flow; a reverse bias and a reverse current are said to be present.

4. The current that flows through the emitter and the collector can be controlled by small changes in the current or voltage at the base; this control permits the transistor to perform operations, such as amplication and switching, which are at the base of several electronic functions. Transistors can be of the NPN or the PNP type, depending on the way in which the *n*- and the *p*-regions are alternated in the device.

5. This characteristic is represented by the redundancy concept in the manufacture of semiconductor devices.

6. Of the standard products, programable logic devices allow personalization by the users only after the device is fully fabricated.

7. Recently, however, the digital TV has been introduced.

8. *Electronics Weekly* (1 January 1985) and 'Motorola' in *Elettronica* (January 1985.

9. *Electronics* (13 January 1983).

10. In addition, they represented a special case of the phenomenon of leadership which so much attracted the attention of Schumpeter.

11. For a discussion, Freeman, Clark and Soete (1982) and Rosenberg and Fritschak (1983).

12. Major and minor innovations are 'a homogeneous phenomenon, the elements of which do not differ from one another except by degree' (Schumpeter, 1951, p. 64, fn. 1). See also Schumpeter 1939, p. 101, footnote 1.

13. After the pioneering work of Usher (1954), see Rosenberg (1976, 1982), and Sahal (1981).

14. Golding (1971), Dummer (1978), US Department of Commerce (1979), Freeman (1982) and Braun and MacDonald (1982).

15. See, for example, Rosenberg (1976), Gold (1981) and Sahal (1981).

16. Memory devices have a product life cycle estimated at less than ten years from prototype to obsolescence (United Nations, 1983, p. 111). During the early period of the industry (late 1950s) transistors had an average life of two years (Tilton, 1971, p. 80).

17. The concept of technological regime has several similarities with the concept of technological guidepost (Sahal, 1981) and of technological paradigm (Dosi, 1982). The former refers to a basic design that guides technological change; the latter refers to a pattern of solution of selected technological problems.

18. See the so-called 'Moore Law' (*Scientific American*, 1977), which states that the number of components per semiconductor device doubles every year.

19. See Nelson (1962) and Levin (1982) for several examples.

20. SIA in United Nations, 1983, p. 280.

21. Golding (1971).

22. Tilton (1971), Rosenberg (1976) and United Nations (1983).

23. For LSI devices of higher complexity and integration, however, the reduction in price due to increases in yield is slowed by the increasing importance of software and automation. For a discussion, see United Nations (1983) and Ernst (1981).

24. *Scientific American* (1977).

25. For the European and the American economies see, for example, Landes (1969) and Rosenberg (1972, 1976).

26. Again, the basic references are Hirschman (1958) and Scherer (1982).
27. For a discussion, see Mowery and Rosenberg in Rosenberg (1982).
28. Mowery (1983) provides a general discussion of the demand pull and technology push interpretations of technological change in the semiconductor industry.
29. For a general, as well as detailed, discussion, see Rosenberg (1982).
30. For a general discussion, see Rosenberg (1972).
31. One should read the fascinating chapters on the applications of microelectronics in electronics final products, in *Scientific American* (1977) and Braun and MacDonald (1982) to get an idea of the innovative processes set in motion by innovations in the semiconductor industry.
32. Control does not necessarily mean financial integration. For a general discussion, see Williamson (1975) and Flaherty (1981).
33. Freeman (1982) discusses these types of uncertainty.
34. Defined as 'Self interest seeking with guile' (Williamson, 1975, p. 9).
35. Even when contracts for R&D results can be successfully established, however, firms in high-technology industries may choose to keep an in-house R&D facility together with a production facility, in order to maintain an in-house technological capability and remain competitive. Mowery (1982) shows that, in general, firms only grant contracts for low-risk, standardized or specialized projects to external R&D facilities.
36. The organizational choices of firms are also affected by the presence of productive, marketing or technological complementarities. These complementarities occur when economies of scope exist; that is, when the cost of producing two products by a single firm is less than the cost of producing them by two separate firms (Baumol, Panzar and Willig, 1982). Economies of scope result when an input is common to two or more products or when indivisibilities occur at the firm level (Baumol, Panzar and Willig, 1982). These indivisibilities may be constituted by 'system' or 'firm'-specific know-how, where 'system' refers to the technology used by the firm (Teece, 1980). When economies of scope are present a firm either diversifies or integrates vertically.
37. The other reasons for vertical integration will be examined in detail in the next chapters, when the specific cases of the vertical integration of firms in Europe, the United States and Japan, will be studied.
38. For an analysis of the Transitron case, see Tilton, 1971.
39. For a more general discussion, see Schakle (1972); High (1983); Langlois, (1983); and Symposium on Uncertainty (1984). Hicks (1979, p. 21) claims that 'when the number of observations is small, the probability calculus is useless.'
40. The major references here are Simon (1957), (1972), and (1979).
41. On this subject, see Rosenbloom and Abernathy (1982), on the American consumer electronics industry.
42. One may find many other significant examples in books such as Tilton (1971); Golding (1971); Wilson, Ashton and Egan (1980); Braun and MacDonald (1982); and in journals such as *Electronics*.
43. It is possible to argue that uncertainty is the main cause of the predictability of behavior among firms (Heiner, 1983).

44. The different attitudes of firms toward major or minor innovations can be very effectively defined by Klein's definition of rationality (Klein, 1977, p. 23): happy warrior, middle-class, accounting, conservation of power.
45. Freeman proposes a wide range of strategies: offensive (aimed at maintaining or consolidating an established position), imitative (aimed at imitating successful innovations), dependent (characterized by a position of dependence with respect to another firm), traditional (with no change in products or processes), and opportunistic or niche (aimed at market niches or at specific consumer needs) (Freeman, 1982).
46. The first group is composed of firms such as Texas Instruments, Motorola, and Mullard (Philips) while the second group includes Plessey and Ferranti.
47. Atkinson and Stiglitz (1969); Rosenberg (1976); Nelson and Winter (1982).
48. Such as managerial talent. At this point it is necessary to remember that firms are also fungible organizations; that is, not entirely specialized in the production in which they are currently engaged. On this issue, see the extensive work of Teubal (1983), and Teece (1980, 1982).
49. For a more general discussion, see Alchian (1950); Day and Groves (1975); Nelson and Winter (1982).
50. For a general discussion, see Arrow (1962); Dasgupta and Stiglitz (1980a, 1980b).
51. See *Business Week* (11 May 1981); *The Economist* (6 June 1981). For a discussion, see Taylor and Silberston (1973); Levin (1982); MacDonald (1982).
52. See Levin (1982).
53. This growth includes some cyclical behavior around the trend: semiconductor recessions occurred in 1960–1, 1966–7, 1970–1, 1974–5, 1981–2.
54. Schumpeter never considered, however, the developments in science and technology as necessary conditions for the technological progressiveness of large firms in specific sectors.
55. See, for example, Levin (1982); Loury (1979); Dasgupta and Stiglitz (1980a, 1980b); Futia (1980); Nelson and Winter (1982); Reinganum (1982).

3 Sector-specific factors in the evolution of the semiconductor industry

Introduction

Using the conceptual framework developed in the previous chapter, it is now possible to lay the ground for an analysis of the evolution of the European semiconductor industry. This analysis will consider the characteristics of technological change, the linkages and interdependencies between the semiconductor industry and the other electronics sectors, the variety of semiconductor firms and the types of industry dynamics in a comparative perspective. In this chapter, some sector-specific factors which account for the evolution of the European semiconductor industry are introduced. These factors, specific to the semiconductor industry, differed across Europe, the United States and Japan. They are divided into supply (Section 1), demand (Section 2) and public policy (Section 3) factors. Section 4 contains some additional observations about linkages and international competition.

1 Supply factors

Supply factors include the organization of R&D and production within firms, the routines and strategies of firms, and the structure of the industry. The organization of R&D and production within firms concerns the choice to integrate vertically not only the R&D and production stages, but also the R&D and production of semiconductors with the R&D and production of electronics final goods and equipment. As discussed in the previous chapter, routines refer to the repetitive pattern of activity within firms, and strategies refer to their long-term plans for survival. Both involve subjective decisions based on limited (bounded) rationality. In addition, firms accumulate firm-specific technological and productive experience with time as a result of past experience, performance, learning-by-doing and learning-by-using. Therefore, both the technological and productive capabilities and the strategies of firms may differ. The structure of an industry, therefore, refers not only to the level of concentration, but also to composition in terms of firms that are similar and that form strategic groups.

Supply factors have differed greatly across Europe and the United States in terms of industry structure. In the European semiconductor industry, most firms are established, vertically integrated receiving tube producers; there have been few new or specialized entrants into the industry. This structure is very similar to that found in Japan. In the United States, by contrast, the structure of the industry consists of a larger variety of firms. There are a number of vertically integrated firms, some of which were former receiving tube producers, and some of which were new electronics producers. There is also a group of specialized (non-vertically integrated) semiconductor firms, most of which were new entrants into the industry.

In assessing the role which these differences have played in the evolution of the semiconductor industry, a general introduction to the function of supply factors is required.

A first 'cut' must distinguish between new and established firms. At a general level, new firms have a positive influence on innovation because of their specific characteristics: speed, flexibility, adaptability and commitment. Schumpeter provides an eloquent discussion of the relationship between new firms and technological change:

> Most new firms are founded with an idea and for a definite purpose . . . . Many of them are, of course, failures from the start . . . . Even in the world of giant firms (trustified capitalism) new ones rise and fall into the background. Innovations still emerge primarily with the young ones . . . [and] threaten the existing structure of their industry or sector . . . because industrial progress comes to the majority of firms existing at a given time as an attack from the outside [Schumpeter, 1939, pp. 94–7].

Evidence from the semiconductor industry (Tilton, 1971; Golding, 1971) indicates that during the late 1950s and the 1960s, the large number of new entrants was a major factor in the fostering of innovation in the American industry, giving it a lead over both the European and the Japanese industries. It should be noted that, in the United States, new entrants were particularly numerous during the periods that followed the introduction of the three major innovations (the transistor, the silicon-planar-integrated circuit and the micro-processor) that changed existing technological regimes. But while during the 1950s, established firms retained their leadership in the industry, during the 1960s and 1970s, new firms took over this leadership.

At a general level, several authors have associated established and successful producers with incremental innovations in established technologies. Schumpeter gave credit to large corporations in trustified capitalism as being the major source of innovation. He added that these firms 'coordinate [the innovation] with [their] existing apparatus, and therefore it [the innovation] need not assert itself by way of a distinct process of competition' (Schumpeter, 1939, p. 96). In industries with high rates of technological change, large corporations remain a major source of innovation. These corporations have large R&D laboratories, and have learned how to deal with change and with new technologies. Schumpeter emphasized, however, that 'old [firms] display, as a rule, symptoms of what is euphemistically called conservatorism (sic)' (Schumpeter, 1939, p. 97). He added that 'the "natural" cause of [death] in the case of firms is precisely their inability to keep up the pace in innovating which they themselves had been instrumental in setting in the time of their vigor' (Schumpeter, 1939, p. 95).

As Rosenberg (1982) and Phillips (1971) have observed in the aircraft industry, moreover, familiarity with a technology more often leads to improvements in, rather than radical departures from, that technology. In addition, success in a technology brings 'an apparent proclivity on the part of once successful manufacturers to remain too long with the basic technology of their original success' (Phillips, 1971, p. 91). Sometimes, in fact, changes are so radical and new technologies so different from existing technologies, that even large and technologically progressive corporations may be caught unprepared to handle such changes, particularly if these firms are successful in existing technologies.

This first cut, however, puts too much emphasis on the innovativeness of new firms. In recent times, in fact, established and successful firms in high-technology industries have increasingly routinized the scanning and search for radically new technological alternatives, and are more prompt to adapt and commit themselves to new technologies that threaten their supremacy in existing technologies. This trend results from the realization by most firms that in these industries change is not the exception, but the rule of life. Therefore, while in most high-technology industries the Schumpeterian remark about the routinization of innovation in the R & D laboratories of large firms in the twentieth century still proves valid, his claim about the continuous displacement of incumbent firms by entering firms must be specified more carefully. While from one side the experience, capability and know-how of established and successful firms may indeed induce them to focus on incremental technological change, to disregard new radical alternatives and (at the organizational level) to resist major technological changes, there is more and more evidence of an adaptive behaviour in incumbent firms and of their successful survival under new conditions. For example, the highly innovative record of established and successful corporations, such as IBM and Texas Instruments, is a consequence of their ability to survive successfully under continuously changing technological conditions. These firms, in fact, have been able to enter new technologies and new markets rapidly and resourcefully.

The tangle about the innovativeness of new or established firms is further complicated in the semiconductor industry by the need to operate a second 'cut' between vertically integrated and merchant producers. At the beginning of the industry this second cut overlapped with the first one. Most established firms, in fact, were also vertically integrated (for example, Bell-ATT, General Electric, Philips, Siemens, Hitachi). Later in the history of the industry, however, this overlapping was no longer present; rather, vertical integration added another element of diversity to the semiconductor industry.

Because of lower user uncertainties,[1] the increased interaction between producers and users and the improved use of information, vertically integrated semiconductor producers are able to focus their innovative efforts on technological areas that are of interest to internal customers. They may, therefore, be slower or less committed to respond to opportunities and stimuli for innovation which come from external end users and which differ from internal ones. In addition, because of lower competitive uncertainty, vertically integrated semiconductor producers may be sheltered from the competitive pressure that is typical of the external market, and may therefore carry on inefficient or noninnovative types of production.[2] Such tendencies, however, may be partially offset by the fact that in the semiconductor industry vertical integration is often tapered (i.e. producers sell part of their production on the external market).

In the following chapters the second 'cut' will be used to group semiconductor firms according to linkage type (internal or external) and to the type of end user market. Throughout this study, semiconductor producers will be grouped into two major groups: merchant firms and vertically integrated firms. Merchant firms sell their products on the open market. They may be part of a large conglomerate or a diversified corporation, but they do not have a productive link with any division (group) within the corporation.[3] Vertically integrated firms, by contrast, are involved in both the R&D and production of the semiconductor components that they require in their final goods. Few vertically integrated firms produce in-house all of the semiconductors needed for internal consumption. Some of these firms (i.e. IBM and ATT) until the early 1980s were captive producers: that is, they sold their semiconductor components only to internal customers. Others had a taper integration because they sold part of their production on the open market. Tables 3.1 and 3.2 list the major firms in each strategic group in Europe, the United States and Japan, both at the beginning of the industry (early 1950s) and in the early 1980s.

Vertically integrated producers in the semiconductor industry are further divided into two groups (see Tables 3.1 and 3.2). The first group includes firms that produced large quantities of receiving tubes before the invention of the transistor. It should be noted that in this first group, European firms produced mainly electronics products for the consumer, telecommunication and industrial markets (see Table 3.3). The second group includes firms that were not established producers of receiving tubes before the invention of the transistor. Many of these producers were specialized in a well-defined range of electronics products, e.g. computers.

A sketch of the major structural changes that have occurred in the European semiconductor industry during the past three decades is indicated in Table 3.4. The European semiconductor industry underwent significant structural changes particularly in France and Great Britain. It should be noted, however, that the division of firms by nationality (in this study, France, West Germany, Great Britain, Italy, the United States and Japan) presents a problem for the classification of two firms: ITT and Philips. Although the central headquarters of ITT

Table 3.1 Strategic groups in the European, Japanese and American semi-conductor industries, early 1950s

1. *Major vertically integrated receiving tube producers*

Europe	United States	Japan
West Germany	Western Electric (ATT)	Nippon Electric
Siemens	RCA	Hitachi
AEG-Telefunken	General Electric	Toshiba
Valvo (Philips)	Westinghouse	Matsushita
The Netherlands	Sylvania	Mitsubishi
Philips	Philco	Kobe Kyogo
France	Raytheon	
Thomson-CSF		
CGE		
Radiotechnique (Philips)		
Great Britain		
GEC		
Mullard (Philips)		
AEI		
English Electric		

2. *Major vertically integrated firms which were not established receiving tube producers*

Europe	United States	Japan
West Germany	Motorola	Sony
SEL-SAF (ITT)	Honeywell	Fujitsu
France	Sperry Rand	Japan Radio
LMT and LLT (ITT)		OKI
Great Britain		
Ferranti		
Lucas		
Plessey		
STC (ITT)		

3. *Major merchant producers*

Europe	United States	Japan
West Germany	Texas Instruments	—
Eberle	Transitron	
Semikron	Fairchild	
Intermetall		
France		
Soral		
Italy		
SGS		

Note: Firms in parentheses refer to parent companies.
Source: Tilton (1971), Golding (1971).

Table 3.2 Strategic groups in the European, Japanese and American semi-conductor industries, early 1980s

1. *Major receiving tubes vertically integrated producers*

Europe	United States	Japan
West Germany	Western Electric (ATT)	Nippon Electric
Siemens	RCA	Hitachi
AEG-Telefunken	Westinghouse	Toshiba
Valvo (Philips)	Signetics (Philips)	Matsushita
The Netherlands	Intersil (GE)	Mitsubishi
Philips (Signetics)		
France		
Thomson-CSF		
EFCIS (Thomson-CSF)		
RTC (Philips)		
Great Britain		
GEC		
Mullard (Philips)		

2. *Major vertically integrated firms which were not pre-established receiving tubes producers*

Europe	United States	Japan
SGS-ATES (STET)	Motorola	Sony
Ferranti	IBM	Fujitsu
Lucas	NCR	Fuji
Plessey	United Technologies	Sanyo
Matra-Harris	Honeywell	Japan Radio
Intermetall (ITT)		OKI
Schlumberger		
ITT (UK)		

3. *Major Merchant Producers*

Europe	United States	Japan
INMOS	Texas Instruments	Kyodo
	National Semiconductors	
	Intel	
	Mostek (United Techn.)	
	Fairchild (Schlumberger)	

Note: Firms in parentheses refer to parent companies. Firms such as United Technologies and Schlumberger are both in the vertically integrated, non-receiving tube producer group and in the merchant producer group (as parent companies). The reason is that they control a merchant producer who partly supplies them with semiconductors.

Source: Company reports; business.

Table 3.3 The specialization of the major European electronics producers in the early 1980s

	Electronic components	Consumer electronics	Telecomm. equipment	Computers	Office equipment	Industrial-Test-medical electronics	Professional electronics	Electrical equipment
Philips	C	A	C		C	B	B	
Siemens	C		B	C		C	C	B
AEG-Telefunken	C	A	B		C			B
Thomson-CSF	B		B	C		B	B	
Olivetti				C	A			
Plessey			A					

Notes: A = sales greater than 30% of total sales
B = sales between 10% and 30% of total sales
C = sales less than 10% of total sales

Source: Company Reports

are located in New York, the American market is not the major concern of the corporation. In the early 1920s, when ATT and Western Electric had to divest themselves of their international operations, ITT purchased them. A subsequent agreement between ATT and ITT gave ITT exclusive use of Western Electric patents outside the United States and Canada; ATT agreed not to compete outside the United States and Canada and ITT agreed not to interfere in ATT's business in the United States.[4] Therefore, ITT's subsidiaries have focused their efforts on the European market. ITT is divided into divisions and world regional areas. Its main European subsidiaries include Standard Electric Lorenz (previously Standard Electric) in West Germany; Standard Telephone and Cable (STC) in Britain;[5] and LMT and LLT in France. Because of the peculiarity that these firms are subsidiaries of an American corporation that has its major interests outside the United States, however, their R&D and manufacturing efforts have been supported and their production has been purchased by various European governments.

Philips, on the other hand, has its central headquarters in Eindhoven, the Netherlands. Because the Dutch market is too small to support large corporations, Philips has become a multinational corporation by taking over existing corporations in several European countries. Philips has a two-dimensional structure (divisions and national organizations) and it has subsidiaries in several European countries: its major subsidiaries include Valvo in West Germany, Mullard in Great Britain, and Radiotechnique Compelec in France. Philips has pursued a policy of strengthening the ties between its subsidiaries and the

Table 3.4 Major structural changes of the main European semiconductor producers

	1950s	1960s	1970s
Netherlands			
Philips	————————————————		
West Germany			
Siemens	————————————————		
Telefunken	————————————— AEG-Telefunken —————		
Valvo (Philips)	————————————————		
Intermetall	——— (Clevite) ——— (ITT) ———		
France			
COSEM (CSF)	⎫		
SESCO (Thomson)	⎬ GE (USA) participation ⎫ SESCOSEM — Thomson-CSF		
Compelec (CGE)	⎭		
Radiotechnique (Philips)	————————————————		
EFCIS	(CEA) — (Thomson-CSF)		
Great Britain			
Mullard (Philips)	———————— ASM ———— Mullard		
GEC	⎫ GEC		
AEI			
EEV-Marconi			
Lucas	⎫ Associate Transistors		
Ferranti			
Plessey			
Italy			
SGS	SGS-Fairchild ——— SGS (Olivetti) (STET) ⎱ SGS-ATES		
ATES (STET)	SGS (STET) ⎰ (STET)		

Note: Firms in parentheses refer to parent companies

domestic markets in which they operate; for these reasons, European national governments have supported Philips' subsidiaries and have purchased their products.

2 Demand factors

Demand factors include the size, growth and sophistication of demand. At a general level, a large size and a high rate of growth (either past or expected) of demand affects the rate of technological change because firms will invest considerable amounts of resources in R&D in order to have large market shares and, therefore, high profits. During the history of the semiconductor industry, the markets for semiconductor devices in the United States and Japan have been consistently larger than those in the various European countries.[6]

In addition, and more importantly, demand factors include the structure of demand in each country. At a general level, demand is not homogeneous either in terms of the products required or in terms of the agent, sector or country of origin. Demand, in fact, may concern either different products within the same industry (in the semiconductor industry, for example, transistors vs. integrated circuits) or the same products with different characteristics (in the semiconductor industry, transistors with different frequencies or integrated circuits with different levels of integration).[7] In this latter case, products may differ in terms of technological attributes, sophistication, complexity, price, reliability, performance or customization. Demand may also come from different agents such as the government, consumers, or firms. Military procurement, for example, puts emphasis on the performance, sophistication and customization of products. Consumers and firms represent a civilian type of demand: they put emphasis on price and reliability, consumers being generally more concerned with price than reliability. In addition, demand coming from firms may be internal to the firm (in-house demand) or may go through the external market (external demand). In-house demand is usually focused on custom and sophisticated products. Finally, overall demand may come from domestic or foreign sources, and industrial demand may come from various industrial sectors.

It is likely that different structures of demand across countries may affect the rate and direction of technological change. Different structures of demand have different 'characteristics' because each demand segment has a different 'weight' within overall demand. Therefore, they may exercise a different pull on innovation.

In the following chapters, demand for semiconductor devices is separated into public procurement and firm demand. Public procurement is divided into military and non-military, while firm demand is divided into four major sectors: consumer goods, computers, telecommunications equipment, and industrial/

professional/instrumentation/medical equipment. The demand coming from each sector is further separated into in-house and external demand.

According to this classification, the structure of demand in Europe differed from that in the United States during the 1950s and 1960s and from that in the United States and Japan during the 1970s. During the 1950s and 1960s, consumer demand was relatively greater in Europe, while military and computer demand was relatively greater in the United States. During the 1970s, on the other hand, consumer and telecommunications demand was relatively higher in Europe, while computer demand was still relatively higher in the United States and computer and consumer demand was relatively higher in Japan.

3 Public policy factors

Public policy factors include the size of public support, the major agency of implementation, the strategy followed and the tools applied. The size of public support concerns the amount of resources devoted by the government to the public procurement of innovative products, to private R&D support, to government conducted R&D and to subsidies for plant and equipment for new products. The agencies are grouped into military and civilian. The type of approach involves such aspects as the goals set, the level of co-ordination and the amount of 'dirigism'.

The goals of policy may concern innovation over the entire technological frontier, innovation in some specific technological areas or the reduction of a technological and productive lag between domestic and foreign firms. The main locus of decisions about such goals, and about the ways to achieve them, may be located within the government or within the firms themselves. In the first case, the government may attempt to direct firms through the use of detailed plans or programmes. In the second case, the government may simply organize or co-ordinate the activity of firms and provide financial support. The scope of policy, and the major agencies involved, will vary across countries.

Finally, the tools used include public procurement, R&D support and government conducted R&D. At a general level, these tools are used by the government for different purposes. Public procurement may stimulate innovation by establishing targets for the products required and by guaranteeing a market for innovative products. By reducing the technological uncertainty connected with the results of the R&D efforts of firms and by reducing the market uncertainty connected with the introduction of innovative products, public procurement affects the profitability and, therefore, the intensity of the R&D efforts of firms. Government R&D support may push innovation by supporting R&D in certain technological areas, by reducing technological uncertainty and by permitting the accumulation of experience in certain technologies. In addition, government R&D support may increase the rate of diffusion of innovations in an industry by obliging firms to disclose their **R&D** results. Finally, government conducted **R&D** may advance basic science and technology and expand the more general scientific and technological infrastructure (including scientific and technological information, design services, technical assistance, etc.). In this way it may reduce the overall technological uncertainty that surrounds firms in the innovation process and contribute to the diffusion of innovation.[8]

During the history of the semiconductor industry, the public policies of the various European countries differed from those of both the American and the Japanese Governments. Until very recently, moreover, the policies of the various European countries were independent of each other and unco-ordinated. Except in part in Great Britain and France, these policies mainly concerned civilian areas. Policy in the United States, on the other hand, was dominated by the military during the 1950s and early 1960s. Finally, the Japanese Government followed a policy of co-ordination and support which targeted specific objectives.

4 Some additional observations about linkages, appropriability, and international competition

In the following chapters, supply, demand and public policy factors will be seen in a dynamic perspective. That is, the analysis will focus on the effects of these factors on the rate and direction of technological change and on the evolution of the industry.

It should be noted that beyond the upstream and downstream linkages between the semiconductor industry and the electronics final goods sectors discussed in the previous chapter, other linkages exist between supply, demand and public policy. First, vertical integration represents a characteristic of supply and, at the same time, a component of demand (in-house demand). Therefore, a single firm may be able to manage both supply and demand internally. Second, public procurement is a component of demand and, at the same time, a tool of public policy. Therefore the government may be able to influence exogenously the level and composition of demand addressed to the semiconductor industry. Vertical integration and public procurement linkages must be added to intersectoral linkages in order to yield a full account of the interrelationships which exist between the semiconductor industry and the rest of the economy.

Supply, demand and public policy factors interact in both articulated and subtle ways with appropriability conditions. As previously mentioned, in the semiconductor industry, patents do not protect the appropriability of the returns on innovations. Appropriability, on the other hand, is highly related to dynamic concepts such as learning curves and lead times.[9] Appropriability, however, can refer not only to the returns from innovations, but also to technological knowledge. In this case, appropriability can vary widely from country

to country: it is possible to argue that low appropriability conditions, and therefore a wide diffusion of technological knowledge among firms in specific technological areas, is the consequence of extensive hiring of R&D and production
employees away from competitors or the result of close co-operation among
firms in basic R&D.

Finally, when international competition is analysed, the degree of protection
of markets and the locational advantages of firms may favour domestic firms
over foreign firms or subsidiaries of multinational corporations.[10] At a general
level, markets may be protected by either tariffs or non-tariff barriers such as
national standards, government R&D and investment subsidies, and public
procurement. If markets were internationally integrated and completely open,
and if firms were rapidly adaptable and situated at the technological frontier,
protection would not be a significant factor, because firms would face a world
demand with the possibility, and the capability of supplying any part of that
demand. However, because firms are evolving and historical organizations, active
in a world of change, protection favours and supports firms located within
particular markets.

In industries such as the semiconductor industry, firms located within a particular market may have specific advantages over foreign firms that are characterized by a high rate of technological change and a demand structure that
differs across countries and that varies over time. In such industries, firms face
uncertainty not only about developments in technology, but also about the
evolution of the specific demand structure in each country. Because firms are
guided by routines and subjective beliefs, their strategies concerning specific
markets with particular demand structures may be favoured by a better understanding of the trend and characteristics of that market. This understanding is
obtained by the firm through direct production in that market: in this case,
firms are better able to interact directly and continuously with customers,
to compete closely with potential rivals and, more generally, to make themselves subject to the stimuli coming from that specific market. This advantage,
which may be called a locational (proximity) advantage, characterizes not only
the semiconductor industry, but also other components, capital goods and consumer goods industries.[11]

In several circumstances, domestic firms are more able than foreign subsidiaries to exploit this locational (proximity) advantage. If the demand structure in the host country is different from that in the source country and if this
structure is changing rapidly, foreign subsidiaries may be slower (or less committed) than domestic firms to grasp these changes in the demand structure for
two reasons, one related to their establishment, the other to their evolution.
First, foreign subsidiaries are set up because the parent company possesses
specific advantages (Dunning, 1979) that may be related to the structure of
demand in the source country. In this case, foreign subsidiaries have advantages
over domestic firms in those specific demand segments which constitute the bulk

of the demand in the source country. Second, once established, subsidiaries usually maintain either a technological dependency or some technological interaction with their parent companies (Caves, 1982). Because of this dependency or interaction, foreign subsidiaries may be more sensitive to changes in the demand structure of the source country than to those of the host country.

The next chapters will examine the effects of supply, demand and public policy factors on the evolution of the European industry. Constant reference will be made to the American and the Japanese semiconductor industries. The analysis will be divided into three periods: the transistor, the integrated circuit and the LSI period.

Notes

1. User uncertainty refers to the likelihood of acceptance of the product and its cost by the user.
2. Competitive uncertainty refers to competition by other producers in terms of innovation and prices.
3. In the rest of this study, the term 'division' will be used as a synonym for 'group', unless otherwise specified.
4. Sampson (1973).
5. Now sold (1984). ITT remains in STC with a 35 per cent share.
6. A European market as such has never existed because national markets have remained separate.
7. This approach is very similar to that of Lancaster (1972 and 1979).
8. For a general discussion, see Pavitt and Walker (1976) and Freeman (1982).
9. This is common to several other industries, as Levin, Klevorick, Nelson and Winter (1984) have emphasized for the American industry.
10. Direct foreign investments in the semiconductor industry have been a major area of analysis. Among the various studies, see Finan (1975) and Flaherty (1983), who provide a detailed investigation of the technology transfer and the international semiconductor industry.
11. This has been described briefly in some of the recent literature concerning multinational enterprises. At the theoretical level, it has been mentioned by Hirsch (1976) and Caves (1971); at the sectoral level, in a number of case studies; see Dunning (1973) and Behrman (1969). However, all these contributions emphasize the need to understand the static characteristics (and not the dynamic changes) of demand in a specific market as the cause of direct foreign investments.

4 The transistor period: the rise and success of the European industry

In this chapter the evolution of the European semiconductor industry during the transistor period will be analysed. The transistor period covers the years between the invention of the transistor (1947) and the introduction of the planar process and the integrated circuit in 1959–61.

A brief review of the invention of the transistor is presented in Section 1. A discussion of the technological regime, the major innovations and the technological trajectories is advanced in Section 2. The history of the evolution of the European semiconductor industry and of its success follows in Section 3. The major factors (supply, demand and public policy) affecting the evolution of the European industry are analysed in Sections 4, 5 and 6. Finally, the evolution of the American and the Japanese industries is discussed in Section 7, by way of comparing them with the European industry.

1 The invention of the transistor

On 23 December 1947, the transistor was invented at Bell Laboratories in the United States. The transistor was a solid-state amplifier which revolutionized the existing electronics industry and officially established the semiconductor industry. This invention was the outcome of the work of three physicists (Bardeen, Brattain and Schockley, each of whom had different areas of expertise), working in a large investment project at Bell (approximately $1m. between 1945 and 1948) in the R&D of solid-state physics and components for telecommunication equipment.[1]

The transistor was discovered by accident (Nelson, 1962), while the research team at Bell was working on finding a type of amplification mechanism (the field effect) (Braun and MacDonald, 1982). This demonstrates the role played by uncertainty in the innovative process, in particular in the early stages of the industry.

Bell's commitment to research in solid-state physics was part of the program which the American industry, left untouched by the war, had been involved in since World War II. After the war not only major corporations, but also the American government, began to finance university research projects on the use of germanium and silicon in radars.

This type of commitment was not possible in any European firm. Most European industries, in contrast to the American industry, had been destroyed during the war and had had to be rebuilt afterwards. In addition, in countries like West Germany, during the process of reconstruction some types of production were precluded and public military procurement was non-existent.

The damage done to the European industry, however, did not mean that this industry was necessarily at a technological (or competitive) disadvantage relative to its American counterpart; following the war, technology was still based on electron tubes, a field in which most European corporations had accumulated experience and capability before the war. This was particularly true for the British, French and Dutch (Philips) industries.[2] Moreover, at the moment the transistor was discovered, research in France and Great Britain on solid-state amplifiers was at an advanced stage. Relying only on public sources of information, in fact, Philips was able to produce a workable transistor only one week after the announcement of the Bell discovery; Thomson-Houston, GEC and STL (ITT) followed shortly thereafter (OECD, 1968).

2 Technological regime, major innovations and technological trajectory

The transistor established a new technological regime. This regime replaced the old one based on electron tubes. It used semiconductor materials such as germanium and silicon, and was based on advances in solid-state physics. It aimed at performing the rectification and the amplification functions.

Some of the features typical of the invention of the transistor were to shape the subsequent technological trajectory of the transistor period. First, science played a very important role in the discovery of the transistor. In fact, the transistor was a scientific discovery. Scientists, rather than practitioners, were the protagonists of the invention because only scientists who had a knowledge of solid-state physics could understand the workings of a solid-state amplifier and could, therefore, develop the transistor. In the years following the invention of the transistor, science continued to play an important, albeit decreasing, role in innovation in the semiconductor industry. Second, interdisciplinary work and organizational structure played an important role in the discovery of the transistor. In fact, interdisciplinary work among people within different areas (physics, chemistry, etc.) was a necessary and vital part of the research that led to this discovery and the decentralized decision-making process proved a viable and successful way of stimulating innovation. This pattern of interdisciplinary work was to continue throughout the evolution of the industry. Third, managerial attitudes and technological beliefs played a role in the discovery of the transistor. The management at Bell Laboratories generally believed in the possibility of a solid-state amplifier,[3] although in several instances the management opposed specific projects or initiative in this field.[4] Even Bell's management, however, did not envision the transistor as the new basic component of the electronics industry. In fact, in the early days of the industry, managers and technicians considered the transistor as a curiosity, partly because it performed less satisfactorily than did receiving tubes. Only later did they see it as a substitute for receiving tubes.

The transistor invented at Bell was a germanium point-contact transistor. It was introduced commercially by Western Electric (ATT subsidiary) in 1951.[5] This transistor however, was unreliable, difficult to manufacture, and unable to handle high frequencies.

The point-contact transistor was followed by the single crystal growing process and the grown-junction transistor. Two types of grown-junction transistors were developed: the germanium-based grown-junction transistor[6] and the silicon-based grown-junction transistor.[7] The former was restricted to low frequencies and was used in early transistor radio sets; the latter was able to function in high temperatures and increased the frequency range of transistors.[8]

The grown-junction transistor was followed by the alloy process, the alloy-junction transistor, the jet-etching process and the surface-barrier transistor.[9] Alloy transistors could operate at a higher frequency and at higher currents than could junction transistors. They also exhibited superior switching properties and were cheaper to produce. They were used in hearing aids, radio sets and in the second generation of computers. With the alloy process, large-scale manufacturing became possible.

Then in 1955–6 three major innovations occurred: the oxide-masking process, the diffusion process and the diffused transistor.[10] The oxide-masking and diffusion processes made the batch production of semiconductors possible; they also lowered the production costs and increased the reliability of semiconductor devices. Diffused transistors, on the other hand, were a product innovation; they had greater power and frequency capabilities than other types of transistor.

The oxide-masking and diffusion processes opened up a completely new phase in the semiconductor industry and constituted the base for several innovations: the drift transistor,[11] the mesa transistor,[12] the post-alloy diffused transistor, and the planar process. In fact, as is shown in the following chapter, the planar process became the standard process for manufacturing semiconductor devices and opened the way to the integrated circuit.

During the transistor period, most process innovations were closely linked to product innovations. At a general level, Abernathy and Utterback (1975 and 1978) claimed that product innovations are more frequent in the early stages of an industry, while process innovations characterize the later stages. The early stages of the semiconductor industry, on the other hand, have been characterized by both product and process innovations. The grown-junction transistor, for example, was introduced together with the grown process, the alloy transistor with the alloy process, the diffused transistor with the diffusion process, the planar transistor with the planar process, the epitaxial transistor with the epitaxial process.

In the transistor period, both technology and production changed considerably. This was related to the very early phase of the history of the industry. In the first half of the 1950s semiconductor technology was fluid and constantly evolving and production processes were undefined and inefficient.[13] Moreover, the life cycle of both products and processes was very short: each new innovation greatly improved existing products and processes.

The technological trajectory of the transistor period was directed towards reliability and performance. Braun and MacDonald (1982) describe this trajectory very effectively. During this period,

> the methods of some companies were embarrassingly empirical . . .[14] and semiconductor manufacture was generally dependent on rudimentary and *ad hoc* principles. . . . The direction of semiconductor development during the early and mid-fifties was determined not so much by the desire to make better devices as by the desire to find better ways of making them. Certainly there was interest in improving characteristics, in creating new types and in extending the applications of semiconductor devices generally, but primary interest was in improving the process by which devices could be made in quantity. The better product was important, but more important was the ability to make devices that were reliable, reproducible and cheap. [Braun and MacDonald, 1982, pp. 66–73.]

This situation of technological fluidity changed, however, in the second half of the 1950s. Following the development of the oxide-masking and diffusion processes, technological change in the semiconductor industry became increasingly cumulative.

3 The evolution of the European industry: a case of success

3.1 *The evolution of the world industry*

During the 1950s, the semiconductor industry remained limited in size, but experienced high rates of growth. Almost non-existent at the beginning of the 1950s, the world semiconductor industry reached sales of $115m. in 1956 and $750m. in 1960 (Freeman, 1965 and Finan, 1975).

Together with these high growth rates, the industry also began to show a sharp sensitivity to the business cycle. In fact, the semiconductor industry experienced a recession in 1960–1. In the United States, for example, unit shipments of transistors and of diodes and rectifiers decreased respectively by 32 per cent and 35 per cent in 1961. Shipments in terms of value, on the other hand, decreased by 1 per cent in 1961 and 3 per cent in 1962 for transistors and by 12 per cent in 1962 for diodes.[15]

The high growth rates of the industry were also accompanied by learning curves phenomena. The prices of diodes and transistors fell during the 1950s as a result of incremental product and process innovations that were made during this period and as a result of learning-by-doing and by-using. Germanium transistors were priced, on average, at $3.56 in 1954 and at $1.70 in 1960, while silicon transistors were priced, on average, at $23.95 in 1954 and at $11.27 in 1960.[16]

The high rates of growth of the semiconductor industry during the 1950s were due to the rapid diffusion of semiconductor devices among users. This diffusion was in turn the result of a growing demand for semiconductors as substitutes for electron tubes, and an additional 'new' demand for semiconductors.

The speed at which semiconductor devices were diffused varied across countries. The United States had a more rapid diffusion process than did either Europe or Japan. In 1961, for example, the ratio value of semiconductor production/ value of electron tube production was 0.74 for the United States, 0.41 for Japan, 0.41 for France, 0.31 for Great Britain and 0.29 for West Germany (Freeman, 1965 and Tilton, 1971).

Within the semiconductor industry, transistors remained less important than diodes and rectifiers during the 1950s. In the United States, for example, the annual factory sales of transistors did not begin to exceed the annual factory sales of diodes and rectifiers until 1960 (Electronics Industry Association, 1979).

During the transistor period, the production and apparent consumption of semiconductors in the United States remained significantly larger than that of all European countries and Japan added together:[17] the figures for West Germany, Great Britain and France were approximately equal, and were lower than those for Japan (see Tables 4.1 and 4.2).

Table 4.1 Production of semiconductors by country, 1958 and 1961 ($ millions)

Country	1958	1961
United States	236	607
Japan	19	78
West Germany	10	30
Great Britain	8	35
France	8	32
Italy	na	na

Note: Rectifiers are excluded from the German data. na = not available.
Source: Freeman (1965)

Despite the production lag, however, European countries were largely self-sufficient in semiconductors during the 1950s and early 1960s. As Figure 4.1 shows, the trade balances of the American and Japanese[18] semiconductor industries were in constant surplus during this period. By contrast, the West German,[19] British and French industries had trade deficits (Tilton, 1971). During the

Table 4.2 Apparent consumption of semiconductors by country, 1956 and 1960 ($ millions)

Country	1956	1960
World	115	750
United States	80	560
Great Britain	2	28
France	2	27
West Germany	3	25
Italy	na	na
Western Europe	9	90
Japan	5	54

Note: na = not available.
Source: Freeman (1965) and Finan (1975).

Table 4.3 Lag of European countries in commercial production of semi-conductors

Country	Average lag in years
Great Britain	1.9–2.2
Japan	2.5
West Germany	2.7
France	2.8
Italy	3.8

Note: This average represents an average time lag in the beginning of the commercial production of a sample selection of new products.
Source: Freeman (1965), p. 64; Tilton (1971), pp. 25, 27.

1950s, however, international trade in semiconductors was not of great importance. In fact, the ratio of exports to the sum of domestic production and imports, although constantly increasing, remained below 10 per cent in the United States. The ratio of imports to the sum of domestic production and imports remained below 20 per cent in Great Britain and France and below 30 per cent in West Germany during the 1950s (Tilton, 1971, p. 44). Therefore, imports constituted only a small part of total consumption in these countries.[20]

During the transistor period the European semiconductor industry performed rather successfully in the world industry. In order to examine the performance of the European industry in detail, it is best to divide the evolution of the industry into two stages: the stage of entry and the beginning of transistor production, and the stage of success in certain types of transistors. The first stage ranged from the early 1950s to the mid-1950s. During this stage, European producers quickly entered the semiconductor industry and

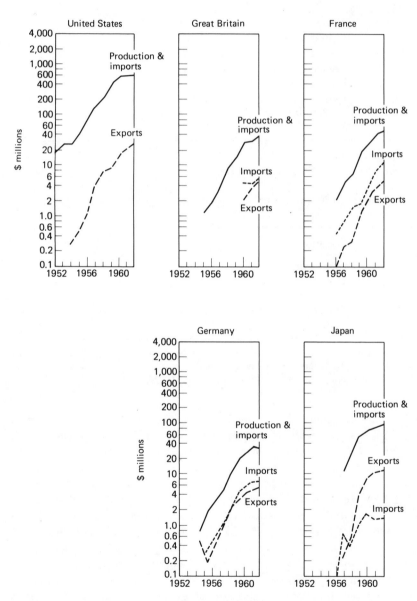

Figure 4.1 Consumption, exports and imports of semiconductors, by country, 1952–1962

rapidly adopted the latest developments of semiconductor technology. The second stage ranged from the mid-1950s to the late 1950s/early 1960s: during this stage, the major European producers became competitive in specific types of products and processes.

3.2 The beginning of the production of transistors in Europe

Vertically integrated receiving tube firms were the first, and the largest, entrants into the European semiconductor industry during the 1950s; merchant semi-conductor producers and American subsidiaries remained small parts of the industry, as Table 4.4 shows.[21] Because the industry was new, entry generally occurred through the introduction of a new production line into existing facilities or through the establishment of a new factory for semiconductor production.

In 1954, for example, Philips built a factory in Nijmegen (the Netherlands) for the production of transistors. At about the same time, Philips tried to pro-duce germanium point-contact transistors and grown-junction transistors. This attempt did not prove successful, however. Philips soon switched to the pro-duction of germanium alloy transistors, the first of which were produced for the general consumer market. Philips also produced germanium diodes and point-contact, small signal diodes; these devices, in fact, eventually constituted a significant part of Philips' semiconductor production in the late 1950s.

Philips diffused the production of germanium alloy transistors and signal diodes throughout its European organization. While specializing in production for specific markets, Philips' subsidiaries in Great Britain, West Germany, France and Switzerland shared a common technological base in semiconductors: in fact, all of the European subsidiaries were producing signal diodes. In addition, Mullard (Great Britain) and Nijmegen (the Netherlands) produced general pur-pose germanium alloy transistors for radios while the Swiss subsidiary produced semiconductors for watches.

Siemens entered transistor production in the same year as did Philips. In 1954, Siemens began the pilot production of point-contact transistors.[22] It soon switched, however, to the production of grown-junction and alloy transistors. Siemens' strategy was different from that of Philips. First, Siemens concentrated its R&D on more areas: for example, semiconductor materials and compounds. Second, Siemens' semiconductor production was concentrated on the internal computer, telecommunication and industrial markets and the external con-sumer market, whereas Philips' semiconductor production was concentrated on the internal consumer and the external computer and industrial markets.[23]

AEG-Telefunken produced fewer semiconductors than did either Philips or Siemens. Many of these devices were used in-house in consumer products such as radios and television receivers. In Ulm, for example, AEG-Telefunken pro-duced germanium diodes and surface transistors for low, intermediate and high frequency bands.

Table 4.4 Major entrants in the European semiconductor industry during the
1950s

	Approximate date of initial commercial production	Product
West Germany		
Vertically integrated producers		
Siemens	1953–4	Transistors
AEG-Telefunken	1953	Transistors
Valvo (Philips)	1953–4	Transistors
Standard Electric-SAF (ITT)	1954	Transistors
Merchant producers		
Intermetal	1954	
Eberle	1959	
Semikron	1960	
Netherlands		
Vertically integrated producers		
Philips	1953–4	Transistors
Great Britain		
Vertically integrated producers		
GEC	1954	Transistors
AEI	1953	Transistors
EE	1955	Transistors
Pye Newmarket	1954	
Lucas	1957	Transistors
Ferranti	1954	Diodes
	1959	Transistors
Plessey	1954	Rectifiers
	1958	Transistors
Westinghouse Brake E.E.	1956	Rectifiers
Mullard (Philips)	1953–4	Transistors
STC (ITT)	1953–4	Transistors
American subsidiaries		
Texas Instruments	1957	Transistors
Clevite	1958	
International Rectifiers	1959	
Emihus (Hughes)	1960	
France		
Vertically integrated producers		
CGE	1954	
Thomson-Houston	1954	
CSF-COSEM	1954	
Radiotechnique (Philips)	1954	
LMT-LTT (ITT)	1954	
Westinghouse-Signal	1954	Rectifiers
Silec	1956	Rectifiers

	Approximate date of initial commercial production	Product
Merchant producers		
Soral	1954	
American subsidiaries		
Texas Instruments	1960	Transistors
Italy		
SGS	1957	Transistors

Note: The names in parentheses indicate parent companies.
Sources: Tilton (1971); Golding (1971); Freeman (1965, 1982).

The major British producers included such firms as Mullard (Philips), AEI, GEC, EEV-Marconi, Lucas, Plessey and Ferranti.[24] Mullard, a large, vertically integrated receiving tube producer, began the commercial production of germanium alloy-junction transistors in 1955 and built a factory in Southampton for the production of semiconductors in 1957. Mullard eventually became the leading British producer of semiconductors and was able to maintain its lead in the electron tube market. The other large British electrical equipment corporations (GEC, EEV-Marconi and AEI) entered the semiconductor industry in the early 1950s. GEC, for example, produced germanium diodes, point-contact transistors and alloy-junction transistors: semiconductors, however, remained only a small part of GEC's overall production.

Lucas, Plessey and Ferranti were among the most successful British entrants into the semiconductor industry during this period. Lucas began producing alloy-junction transistors and diffused rectifiers in 1957 and high-voltage transistors in 1960. But because Lucas was a large vehicle equipment producer, it manufactured silicon, rather than germanium, devices; silicon, in fact, was better suited to the high temperatures that characterized internal combustion engines. As a result, the Lucas group used a large share of its semiconductor production in-house. Plessey received some military contracts for work on silicon in the early 1950s. It began to manufacture silicon alloy rectifiers in 1954. Then, in 1957, Plessey established a joint venture with Philco (Plessey controlled 51 per cent of the equity) for manufacturing a Philco line of transistors; this production began at Swindon in 1958.[25] Ferranti, on the other hand, concentrated its efforts on silicon: in 1954, it began producing silicon diffused transistors.

Finally, some British semiconductor producers specialized in the production of power semiconductor devices, part of which they used in-house. Westinghouse Brake and Signal, for example, was one of the main producers in this line; it began manufacturing germanium rectifiers in 1956 and germanium thyristors in 1958.

In France, both COSEM (CSF) and Thomson-Houston manufactured semi-conductors for in-house use as well as for sale to the military. Most of the other French semiconductor producers (LMT-LTT—an ITT subsidiary—Campagnie des Freins et Signaux/Westinghouse and CGE) were specialized in power devices. Silec, on the other hand, manufactured diodes and rectifiers.

In Italy, SGS was the only major producer to enter the semiconductor industry. It produced germanium transistors.[26]

3.3 *The entry of merchant producers*

Few merchant producers entered the European semiconductor industry during the 1950s. The major explanations for this lull (some of which have been advanced previously) include the absence of the spin-off and mobility of per-sonnel among firms which characterized the American industry, the reduced size of the European electronics market during the 1950s (mainly because of reduced demand on the part of the military and computer industry), and the concentration of government R&D and production refinement support in a few established firms. In Britain the only important merchant producers to enter the industry were Semitron, a spin-off from Plessey, and Brush. In Germany Intermetal was formed in 1952 as a merchant producer but was taken over by Clevite in 1955. In addition, Eberle and Semikron, spin-offs respectively of Intermetal and Siemens, began semiconductor production around 1960. Finally, in France, only Soral, a small specialized firm, began producing semiconductors in the early 1950s (Tilton, 1971).

3.4 *R&D policy and the transfer of technology between American and European producers*

During the first half of the 1950s the resources invested by most European corporations in R&D allowed them to keep up with the technological and commercial capabilities of the leaders in the semiconductor industry. Philips, for example, had a corporate research laboratory—the Naturkundig Laboratory, located in the Netherlands; this laboratory was devoted to basic research and was traditionally involved in serious scientific work. Siemens also had laboratories devoted to basic research in Siemens & Halske and Siemens Schuckertwerke. In the early 1950s, in fact, Siemens made a major effort in research on semi-conductor materials: the Erlangen Siemens group focused its efforts on III-V compounds (Siemens-Berichte), while the higher power devices group in Pretz-feld worked on germanium, silicon and III-V compounds in order to find materials better suited for power devices. In 1953–4 this latter group developed a process for the preparation of high purity silicon—the zone-refining process.

These firms also profited from the fact that, during this period, appropriability conditions in the semiconductor industry were low because all interested firms could easily obtain basic transistor patents; Bell controlled all of the major patents (the point-contact transistor, the grown-junction transistor, the zone refining process, the oxide-masking and diffusion processes) and decided to follow a liberal policy of making these patents available to American, European and Japanese corporations (Tilton, 1971; Levin, 1982). Bell's choice was also adopted by other producers within the industry.[27] As a result, technology was easily transferred from American to European corporations.

Because of the limited capability of written communication, however, this transfer of technology between American and European corporations during the 1950s also occurred through the movement of personnel from the licensee to the licensor (Golding, 1971). This movement involved mainly American and European vertically integrated corporations; several agreements between European and American corporations in fact involved firms which had shared basic patents in the earlier receiving tube oligopoly.

It should be noted that technology was transferred from firms in the United States to similar firms in Europe: several European producers asked for licences from corporations that either had similar experience or structural character-istics.[28] During the 1950s, for example, Philips received licences from Bell and RCA, Siemens from Bell, AEG-Telefunken from General Electric, AEI and GEC from Bell and General Electric, English Electric and Westinghouse Brake from Bell and RCA, Lucas and STC from Bell, Plessey from Philco and General Instruments, Pye from Bell and General Instruments,[29] Cosem from Bell and RCA, Thomson-Houston and SGS from General Electric.[30]

The transfer of technology from American merchant producers to European vertically integrated receiving tube producers was less intense than that between vertically integrated receiving tube producers. At the end of the decade, for example, Texas Instruments had licensed only ITT and Philips (Tilton, 1971); Texas Instruments, in fact, preferred to make direct investments in Europe rather than to license European firms.[31]

3.5 The peak of competitiveness for the major European producers during the second half of the 1950s

During the second half of the 1950s some European semiconductor producers (mainly Philips and Siemens) reached a highly competitive and innovative position in the European market. These European corporations were highly profitable on the external European semiconductor market. However, semi-conductors continued to be regarded in the overall strategy of these vertically integrated producers, as components of larger systems, and as means to increase the competitiveness of their electronic final goods, rather than as independent products.

During this period, Philips and Siemens had the largest market shares and

and were the largest patent holders in the European semiconductor industry (see Tables 4.5, 4.6 and 4.7): they occupied these leading positions partly as a result of the efficient processes which they had developed for manufacturing standard devices. Philips, for example had developed efficient production methods for the production of standard germanium alloy transistors for the external computer market as well as for internal use. Philips' alloy-junction transistors were priced low, had relatively high switching and frequency capabilities and were consequently extremely successful on the European market.

Table 4.5 Market shares of major firms in the British semiconductor market, 1958 and 1962

	Transistor market (%): 1958
Mullard (Philips)	55
AEI	31
GEC	3
Other	11
Total	100
	Semiconductor market (%): 1962
ASM (Mullard-GEC)	49
Texas Instruments	13
Ferranti	10
AEI	7
Westinghouse Brake	5
STC	2
International Rectifier	1
Hughes	1
Other	12
Total	100

Note: Internal sales are included.
Source: Golding (1971), p. 179.

Philips and Siemens, however, were also successful because of the incremental innovations which they had made in two of Europe's end-use markets: the consumer and industrial markets. These innovations included the post alloy diffused transistor, an improved mesa transistor for the consumer market, and an improved silicon controlled rectifier for the power device market. These innovations were based on a common technological belief: both Philips and Siemens thought that, because it performed better at higher temperatures, silicon was more suitable for power devices and that, because of the higher mobility of its electrons, germanium was more suitable for consumer devices. They also thought that technological change in semiconductors had achieved

Table 4.6 British semiconductor patents awarded to major firms by country, 1952–1960

United States	262	Great Britain	213
Western Electric	63	STL	73
		GEC	41
West Germany	96	BTH	28
Siemens	65	NDRC	10
AEG-Telefunken	31	Westinghouse Brake	9
		AEI	9
Philips-Mullard	76	SES	9

Source: Golding (1971)

Table 4.7 French semiconductor patents awarded to European firms, 1954–1960

Siemens	145
Philips (Mullard, Valvo, Radiotechnique)	141
Thomson-Houston	122
ITT (STC, Standard Electric, LMT-LTT)	49
AEG-Telefunken	42
GEC	30
Westinghouse Brake	22
CSF	19

Source: Tilton (1971).

its second major innovation with the diffusion process and that any other major innovations were unlikely.[32]

The incremental innovations which Siemens and Philips made in consumer devices were directed at fulfilling a well-defined need. Radio producers had for some time been demanding devices that would be capable of handling both the short wave and the VHF range of the frequency band. Some of these producers were in-house customers, as in the case of Philips, while others were external customers, as in the case of Grundig with Siemens. Soon after a symposium on the oxide-masking and diffusion processes held by Bell in 1956, and in which several European firms participated (Tilton, 1971), Philips and Siemens envisioned the possibility of producing diffused types of transistors for the high frequency range and, thus, of satisfying the demand by radio producers. These firms had a choice between two types of transistors, both of which were incremental innovations over the Bell process and both of which had been developed by American firms: the RCA germanium drift transistor and the Texas Instruments and Motorola germanium mesa transistor.

Siemens chose to produce the mesa transistor, considered to be the most advanced and reliable transistor available in the 1950s. Siemens' mesa transistors were able to handle frequencies that ranged between 100MH and 1GH (Siemens

Zeitschrift, 1961, 1962). They were manufactured on a large scale in the early 1960s and sold primarily on the external consumer market.

Philips, however, decided to follow another route. In 1957, after the second Bell Symposium, the radio group at Philips placed great pressure on the management board and on semiconductor operations to begin the in-house production of transistors with greater frequency ranges. The semiconductor operation was ordered to run pilot lines for both the mesa and the drift transistors. Simultaneously, however, Philips' semiconductor operation decided to carry out its own independent research activity; it developed the post alloy diffused transistor as a result of this research.[33] Post alloy diffused transistors could be manufactured cheaply and had initial success. These devices were ready for marketing in 1958-9 and were eventually chosen by the corporation for in-house manufacture. They were used in consumer goods, namely radio and television receivers and, in the early 1960s, were produced in Philips' plants all over Europe. The success of post alloy diffused transistors did not last very long, however: they were hard to control and difficult to master. Valvo (Philips' subsidiary in Germany) in fact moved to the production of mesa transistors in order to standardize its devices with those of Siemens. In addition, Philips chose to use the mesa, rather than the post alloy diffused transistor, as VHF television amplifiers.

The second group of incremental innovations made by Siemens and Philips concerned power devices. The development and production of such devices offered many advantages to these corporations. First, corporations such as Philips and Siemens, which had accumulated experience in electrical technology and were competitive in the production of electrical goods, found the power semiconductor device market attractive and well suited to their capabilities. Philips, in fact, was a successful producer of power devices. Philips' interest in the production of power devices resulted from the incapacity of small signal diodes to handle strong currents and high voltages: these devices were not suitable as rectifiers in Philips' television receivers. In 1957 Philips began the development of medium-power alloy silicon rectifiers that went into commercial production in 1958. These rectifiers were initially used in-house in television receivers and later, as Philips became a major producer of these devices, were sold on the external market for telecommunication and industrial uses.

Siemens, a leading producer of electrical engineering equipment, chose silicon as the semiconductor material to be used in power devices. In 1957 it produced silicon rectifiers for locomotives for the Deutsche Bundesbahn and silicon controlled rectifiers for rolling mills or conveying systems (Von Weiher and Goetzeler, 1984). It also produced silicon transistor power devices.[34]

The rest of the European industry consisted of two categories of firms: those that remained marginally involved in semiconductor production (e.g. AEG-Telefunken) and those that slowly pulled out from semiconductor production.

In West Germany, AEG-Telefunken's production of semiconductors was focused on power devices for the industrial market and on diodes and transistors for the consumer market. Among other devices, AEG-Telefunken produced germanium drift transistors that were used in high frequency ranges.

In Britain, several of Philips's and Siemens's competitors suffered major setbacks during the 1950s.[35] GEC lost competitiveness in semiconductor production and, as a consequence, merged with Mullard to form a new company, Associated Semiconductor Manufacturers (ASM) in 1962; Mullard controlled two-thirds of the stock and GEC one-third. *De facto* this meant that GEC exited from the semiconductor industry. AEI and Plessey, two other English firms, also lost competitiveness in semiconductors during the 1950s. After it was pushed out of the consumer market by Mullard's production of post alloy diffused transistors, AEI slowly transferred to the power field. Plessey, on the other hand, continued to produce Philco's surface barrier transistors; this production was unsuccessful, however, and in 1961 Plessey was forced to buy Philco's share of the joint venture. Finally, EEV-Marconi and STC chose to specialize in the production of devices for the telecommunications market. EEV-Marconi manufactured transistors for telecommunications equipment producers that lacked an in-house semiconductor facility (ATE and Ericsson): in 1958, however, EEV-Marconi, ATE and Ericsson founded Associated Transistors, a new firm that produced semiconductors for the telecommunications equipment produced by these firms. STC, on the other hand, produced high-reliability transistors for the telecommunications market, some of which were used by the ITT group.

In France no market leader arose to threaten the European leadership of Philips and Siemens in the semiconductor market. GEC formed a French subsidiary, Compelec, to produce semiconductors, but its production was small and concentrated in the power area. Cosem, a subsidiary of CSF, never attained large dimensions. In addition, 49 per cent of the stock of Sesco, created by Thomson-Houston in 1961, was controlled by General Electric.

Finally, in Italy, SGS's production of semiconductors remained limited and was consistently unprofitable.

By the end of the 1950s and the beginning of the 1960s, therefore, European markets were largely self-sufficient. The consumption of devices imported from the United States or produced by American subsidiaries in Europe was, in effect, limited during the 1950s by the fact that European electronics and semiconductor markets were small and European corporations remained internationally competitive in semiconductor production. In fact, only one major American producer (Texas Instruments) entered the European market through foreign direct investments; Texas Instruments made its first direct foreign investment in Europe in 1957, aimed at the production of a specific device for the British military market, the silicon grown-junction transistor. This investment was favoured by the British military who were concerned with the high level of imports of silicon transistors for military requirements (Golding, 1971). In 1960 Texas Instruments also opened a subsidiary in France.

Other American firms, such as International Rectifier and Hughes, began producing semiconductor components in Europe during the 1950s and early 1960s. International Rectifier set up a joint venture with Lancashire Dynamo in Britain in 1959 for the production of rectifiers for in-house use,[36] and Emihus (a subsidiary of Hughes) began manufacturing diodes in Scotland in 1960. These firms, however, did not have a major impact on the European market. In addition, in the early 1950s, IBM was present in Europe with subsidiaries that produced office equipment. However, IBM did not begin the mass production of semiconductors in Europe until the early 1960s.

Table 4.8 Distribution of United States semiconductors and transistors sales by end use, 1960 and 1963

	Total ($ millions)	Percentage of total market				
		Military	Computer	Consumer	Communi-cation	Industry
1960 Semi-conductors	560	50	30	5	15	
1963 Transistors	252.3	47	17	16	6	14

Sources: Semiconductors, Finan (1975). Transistors, E. Ney Dodson III, in Braun and MacDonald (1982), p. 80.

4 Supply factors: the success of vertically integrated producers

4.1 *Technological capabilities and firm strategies*

During the transistor period, several of the supply factors discussed in the previous chapters had a major influence over the rate and direction of technological change and the competitiveness of European producers: among these factors, technological capabilities, vertical integration and firm strategies were particularly important.

Previously accumulated technological capabilities in electron tube technology allowed European producers to enter the semiconductor industry successfully. European producers had been successfully producing electron tubes for several years and had accumulated production experience, marketing knowledge and technological capability which proved useful for beginning semiconductor production. During the 1950s, the management of several European corporations became convinced that some capability and experience in the R&D and manufacture of semiconductors would enhance the competitiveness of their electronics final products. They realized, in fact, that semiconductors could be used as inputs in almost all electronics final goods, and that these devices would eventually play a key role in several products.[37] Most of these technological beliefs were not well defined; many, in fact, remained mere intuitions. The semiconductor industry was too young and the research and experience in semiconductor technology too limited in the early 1950s to permit any detailed analysis of the technological impact of semiconductor technology on downstream applications.

These technological beliefs did not affect the organizational 'dependency' of semiconductor operations on the final product divisions. The beliefs of the top management of these firms were concerned only with the presence or absence of semiconductor activity within the corporation, not with this activity's relationship with other divisions within the firm.

These technological beliefs, however, were an important factor in the decision of firms such as Philips and Siemens to enter the semiconductor industry. In the early 1950s Philips's management board was strongly convinced that transistors would become major components in the various electronics final goods that were produced by the corporation, particularly consumer goods. The management board of Siemens had similar beliefs; it was convinced that transistors and, more generally, semiconductors, would become key components in their telecommunication and industrial products.[38] As a result, both of these firms moved into the production of semiconductors.

These technological beliefs were also a major factor in the later commitment of Philips and Siemens to the post alloy diffused transistor and the mesa transistor. As a consequence of these beliefs, Philips innovated incrementally with the post alloy diffused transistor and Siemens with the mesa transistor.

4.2 *The organization of R&D and production*

During the transistor period, firms in Europe were involved in both the R&D and production of semiconductors, in part because these firms were, by tradition, involved in the R&D and production of products and, in part, because no specialized R&D firms existed at that time. Moreover, although this R&D activity had to be carried out by a team of specialists, it did not require a prohibitive amount of resources from existing semiconductor producers.

During the second half of the 1950s, however, because of the increase in the kinds of transistors offered on the market, corporations did not find it feasible to produce the entire range of transistors. Rather, these corporations concentrated their in-house production on types of components used in greater quantities in their electronics final goods and on devices that were considered vital for the production of their electronics final goods. This was the case with regard to European firms such as Philips, Siemens and AEG-Telefunken. During the 1950s and 1960s the share of semiconductors which these corporations purchased from external sources increased as the number of types of semiconductor devices increased.[39]

In the following sections, the major factors that induced electrical final goods producers to integrate upstream into the production of semiconductor components are examined. These factors include: the tradition of vertical integration among large electrical final goods producers; the presence of economies of scope in the production of receiving tubes and transistors; the technological beliefs of electrical final goods producers about transistors and the need for a secure supply of transistors.

4.3 *The incentives to integrate vertically: tradition and economies of scope*

Several large electrical final goods producers were 'by tradition' integrated into the production of electrical and electronic components, and especially receiving tubes: in Europe this was true for Philips, Siemens and AEG-Telefunken, among others. Because the transistor was considered a substitute for receiving tubes in certain applications (e.g. hearing aids) and because vertical integration into diodes and receiving tubes was a profitable and successful strategy,[40] these firms also decided to begin producing transistors. By doing so, they could also reap the economies of scope which derived from their production of receiving tubes.

Economies of scope existed for two major reasons. First, semiconductors, like electronic tubes, were components, and a similarity of functions existed betweeen certain semiconductors and certain electron tubes (for example, between transistors and receiving tubes). This similarity meant that electron tube producers had a greater knowledge of the possible uses of semiconductor devices than had inexperienced, new entrants.[41] Second, electrical firms already had the human resources necessary for semiconductor activity: in fact, at the beginning of the industry when there were still few solid-state engineers, these firms employed a large number of electrical engineers, physicists and chemists. It was the physicists and chemists, in fact, who managed the R&D and production facilities of electrical components.

A close examination of the beginning of the manufacture of diodes and transistors in several European firms provides a good illustration of the ways in which economies of scope influenced the evolution of the semiconductor industry.[42] In the early 1950s, Philips' laboratory began, independently of corporate decisions, to undertake R&D in semiconductors because it possessed technological capabilities in this field. This R&D was successful and, as a result, Philips began to manufacture germanium diodes and germanium point-contact transistors.

Siemens, on the other hand, followed a more hierarchical procedure: its laboratory did not move independently into R&D in semiconductors. Rather, it first obtained recognition from the management board that there was a need for such R&D. However, the electrical capabilities of Siemens also played an important role in determining its entry into the new industry. In fact, its experience and previous success in the R&D and production of electron tubes, capacitors and resistors, allowed Siemens to enter semiconductor production rapidly and with a solid technological base.

The same is true for AEG-Telefunken, GEC, Thomson-Houston, STC, and other European electrical firms. These firms began producing point-contact diodes and transistors in the early 1950s, not only because their R&D efforts were targeted specifically at semiconductors and because they were competitive producers of germanium diodes, but also because they had accumulated experience in electrical technology.

4.4 *The incentives to integrate vertically: security of supply and transactional factors*

Given the importance of semiconductors for electronics final goods production, a strong incentive for semiconductor users to integrate upstream was provided by the fear of possible component shortages. At a general level, such shortages are more likely to occur when input markets are small rather than when they are large and developed. Yet, these shortages become increasingly likely when the inputs industry is at an initial stage of development[43] and is characterized by disequilibrium, uncertainty and change. Under such conditions, input users will be induced to integrate vertically: the pressure of the growing demand of inputs coupled with the existence of a limited number of suppliers will induce existing users to integrate upstream because the availability of inputs and delivery times in the new and growing inputs industry is subject to a high degree of uncertainty.[44] In addition, the transactions of a relevant amount of inputs, when overall input supply is limited and overall input demand is growing, may have a high cost because it may be characterized by idiosyncratic exchanges and information impactedness.

In the European semiconductor industry, these transactional factors provided an additional incentive for upstream vertical integration into semiconductor production. During the early 1950s the semiconductor industry was new and small, though the diffusion of semiconductors among users was growing rapidly, not only through a substitution process, but also through the pervasiveness of semiconductor uses. Moreover, there was no merchant semiconductor industry yet in existence in Europe, while one was just beginning in the United States. If European electronics final goods producers had purchased semiconductors from the relatively few American merchant producers that existed, they would (theoretically) have run the risk discussed above. If these producers had purchased semiconductors from other European or American vertically integrated receiving tubes producers, on the other hand, they would have been directly dependent on their competitors at the final product level for supplies of semiconductors. Therefore, these European firms were induced to integrate vertically upstream into the production of semiconductors.

4.5 The consequences of vertical integration[45]

During the transistor period, the routines of vertically integrated producers had well-defined characteristics. While the routines of merchant producers were in most cases new and based on semiconductor technology, the routines of most vertically integrated receiving tube producers, particularly European producers, continued to be based on electrical technology. The managers of these corporations were, in several cases, electrical engineers and the management of the component divisions within these corporations had experience in either electrical or electron tube technology; very few managers had a deep understanding of solid-state technology or a good grasp of its future possibilities. They therefore initially interpreted semiconductor technology as an improvement over electron tube technology, and they considered transistors as devices which had a wider use than receiving tubes, rather than as devices which were completely new.

These views were reinforced by two situations, one at the components division level and one at the corporate level. At the components division level, the beginning of semiconductor activity represented a threat to the dominant position occupied by electron tubes. As a consequence, there was considerable opposition to the extension of solid-state activity within these divisions, especially at the beginning of the industry. Even when semiconductors were produced in plants separated from the rest of the component facilities, rivalries between the electron tube group and the semiconductor group existed. In addition, electrical engineers in other divisions or on the management boards of some of these corporations were more familiar with, and consequently more favourable to, projects and activities involving either electrical or electron tube technology. It should be noted that, at the corporate level, an understanding of the new semiconductor technology by management was hampered by the fact that, because it managed large diversified corporations, management was not closely involved in semiconductor activity.

During the 1950s, the relationship between the producers and users of semiconductors within most European vertically integrated corporations remained one of 'dependency'. The management of the semiconductor operations was relatively free to choose the best policy for the R&D and the manufacture of semiconductors, but only within well-defined limits. These policies could not exist in opposition to the overall strategy of the corporation which was designed to enhance the competitiveness of electrical and electronic final goods. In addition, strategic choices regarding the production of final goods and involving either the in-house production or the external purchase of semiconductors were taken unilaterally without the agreement of the semiconductor divisions.

Finally, within vertically integrated firms, the production of final goods was more flexible and more responsive to external circumstances than was the production of semiconductors. Particularly when most of a firm's semiconductor production was used in-house, adaptation in semiconductor production was provoked more by stimuli coming from the final goods division than by stimuli coming from the external market.[46]

5 Demand factors: the relevance of the consumer market

During the transistor period, the consumer market constituted the largest market for semiconductor devices in Europe, as Table 4.9 shows. Only in Britain was the military market more important than the consumer market. In West Germany, France and Italy, the consumer market far outweighed the importance of the other markets.[47]

Table 4.9 Distribution of European semiconductor sales by end use, 1963

	Total ($ millions)	Percentage of total market		
		Military	Industrial	Consumer
Great Britain	60	50	33	17
West Germany	36	8	36	56
France	34	29	35	35
Italy	12	8	42	50

Source: Lewicki (1966), p. 480.

The different importance of the various markets for semiconductor devices derived from the different importance of the production of electronics final goods in the various countries and its linkages and interdependencies with the semiconductor industry. The quantitative linkages are given in Table 4.10. As can be observed, the consumer market was the most electron tube intensive and the least semiconductor intensive of all the electronics final goods.

The importance of the consumer market for the European semiconductor industry derives from the absolute size of electronics consumer goods production in Europe. European firms, in fact, produced large amounts of, and remained competitive in, electronics consumer goods during the 1950s. Therefore, they created a large demand for semiconductor devices. Part of this demand was internal. In fact, many European corporations which produced electronic consumer goods were vertically integrated corporations. They had a lot of experience in the production of electrical and electronic consumer goods, which originated from their involvement in the manufacture of electrical domestic goods and radio receivers. During the 1950s most of these European corporations manufactured transistor radios and black and white television receivers. In the production of these consumer goods, these corporations used part of

Table 4.10 Semiconductor consumption vs. electron tube consumption
in major electronic equipment industries, 1963

Consumption by manufacturers of	Semiconductors (% of all materials)	Electron tubes (% of all materials)
Electronic computers & calculators	12	1
Consumer electronic equipment	2	21
Telephone & telegraph equipment	7	2
Military & industrial equipment	4	3
Electrical test & measuring instruments	4	5
Engineering & scientific instruments	2	1
Average (excluding telephone & telegraph)	5	7

Source: US Department of Commerce (1979), p. 34.

their in-house production of diodes and transistors; this is true, for example, for Philips, the largest European electronic consumer goods producer, and for AEG-Telefunken.

The rest of the demand for semiconductors which derived from electronics consumer goods producers was external. Therefore, some European semiconductor producers sold most of their production of germanium transistors on the external market; Siemens, for example, sold these devices to Blaupunkt and Grundig,[48] which used these transistors in automobile radio receivers and in tuners.[49]

Because of the importance of consumer demand (both internal and external), European semiconductor producers focused their R&D and production in specific directions. Philips and Siemens, for example, focused their R&D and production on transistors that were capable of handling the high frequency ranges that were required by the demand of radio producers.

In consequence, European producers were successful and innovative in semiconductor services for the consumer market. As shown above, Philips introduced the post alloy diffused transistor while Siemens introduced an improved mesa transistor (that had earlier been introduced by Texas Instruments).

Apart from the consumer market, the other major European market for semiconductors was the industrial one, in particular the power device market. This market was composed of the demand coming from corporations such as Siemens,

AEG-Telefunken, Elliott-Automation, Ferranti, English-Electric, AEI, CAE, Thomson-Houston (Freeman, 1965).

The industrial market, in particular the power device market, affected the direction of technological change in the European semiconductor industry differently than did the consumer market. The market for power devices was less technologically dynamic than the market for small-signal transistors. It was limited to small quantities of each type of device, gave a premium to extreme reliability and was characterized by high prices of the devices. Several European corporations targeted this device on the grounds that it was less volatile than the small signal transistor market, and more attractive in terms of unitary profits. Philips, for example, was a very successful producer of power devices for the external market and, in part, for the in-house consumer market.

European vertically integrated corporations were at even more of an advantage than non-vertically integrated corporations in the power market. These corporations could exploit the advantage of a close communication between producers and users. This close communication was required in the power device market because detailed specifications of the properties and applications of these devices were necessary for their production and use, and the testing of the devices constituted a large part of the total cost of production (Golding, 1971). Therefore, European corporations such as Siemens (vertically integrated into industrial equipment) innovated incrementally in silicon controlled rectifiers, exploiting these communication and test advantages.

Contrary to the consumer and the power device market, public procurement of semiconductor devices had a limited influence on the direction of technological change in the European industry. Public procurement in most European countries was small in absolute size.[50] Only in Britain was the relative share of public procurement over end user markets high,[51] but public procurement did not exert a pull on innovation. In fact, British public procurement followed a low-risk policy of purchasing only established components which had already been successfully launched on the American market by American firms. Moreover, in the early 1960s it began to buy components from American producers for prototype applications. Finally, it did not give extensive support to integration of components into military equipment. As a result, the time lag between the contract and the production of semiconductor devices in Britain increased (Golding, 1971). In the French case, most of the devices required by the military were purchased from two major firms: CSF and Thomson-Houston (Zysman, 1977). In the German case, military expenditures during the 1950s were very limited during the period of post-war reconstruction. However, none of the government's policies, even after this period, were directed toward the semiconductor industry.[52] Finally, in all European countries, the military perpetuated the existing market structure by purchasing semiconductor devices from existing firms.

As with public procurement, so too the computer market had only a limited influence on the rate and direction of technological change in the European semiconductor industry. The reasons for this limited influence concern the size of the market, the lack of success of European firms and the lack of public policy.

The size of the European computer market remained limited, particularly when compared to the size of the American market. Table 4.11 shows that in 1959, in the United States, 2,034 computers were installed, while in Europe only 265 were installed.

Table 4.11 Size of the computer market in the United States and Europe, 1959

	Number of machines installed	Relative size (USA = 100)
United States	2,034	100
EEC and Great Britain	265	13

Source: Diebold, in OECD (1969).

In addition, most European producers were not successful in the commercial production of computers. Indeed, during the pioneering stage of the industry European producers were at the technological frontier and their technological capabilities were similar to those of American producers.[53] This situation is similar to that of the electronics component industry before the invention of the transistor. Later, in the 1950s, however, most European producers were not very successful in the commercial production of computers. Siemens, for example, entered the commercial production of computers in order to enhance the competitiveness of its final products.[54] But it was not able to keep pace with the emerging market for business applications.[55] This was also true for Telefunken and Standard Electric.[56] Bull, on the other hand, had success with medium-size computers but not with large computers, partly because it began to use computers so late.[57] British producers were numerous, but equally unsuccessful.[58] Finally, in Italy Olivetti produced some large computers, but it lacked funds and therefore eventually had to sell its electronics division to General Electric.[59]

In addition, during the 1950s, European public policy towards the computer industry was almost non-existent. European governments provided their domestic computer industries with little procurement and little R&D support. As will be shown later, this situation radically differed from the situation of the American computer industry, which was heavily supported by the government during the 1950s.

As a result of these developments, when the European computer market finally did materialize toward the end of the 1950s, American corporations were able to control a large share of this market. They did so either through exports or through their European subsidiaries, as Table 4.12 shows. Therefore,

Table 4.12 Market shares of computer producers, by country, 1959 and 1962

	Percentage of total number of computers installed 1959[a]	1962[a]	Percentage of value of computers installed 1959[b]
United States			
IBM	65.8		
Sperry Rand	8.7		
West Germany			
IBM	55.6	62.2	73.1
Siemens	6.8	na	13.3
Zuse	29.5	11.7	5.3
UNIVAC	3.4	na	3.9
Great Britain			
ICL		51.0	
IBM		17.9	
NCR Elliot		22.1	
France			
Bull/GE		49.1	
IBM		48.8	

Note: na = not available.
Sources: [a]OECD (1969); [b]Diebold in Kloten (1976).

the demand that came from computer producers located in Europe for semi-conductors produced by European firms was limited for several reasons. First, the demand for semiconductors which came from American subsidiaries in Europe was largely directed toward American semiconductor producers.[60] Second, several European computer producers, such as Siemens and Telefunken, produced a share of the semiconductors that they required in-house.[61] However, because their production of computers was limited, these European producers required only a small number of semiconductors. Third, most of the non-vertically integrated computer producers, such as Bull, ICT and Olivetti, purchased semiconductors on the external market from both European and American producers.[62]

During the transistor period, the telecommunication equipment end-user market also had little influence on the rate and direction of technological change in the European semiconductor industry. Although the discovery of the transistor took place in a firm that produced telecommunication equipment (ATT-Bell-Western Electric), transistors were not widely used in telecommunication equipment. As Table 4.8 shows, in 1963 only 6 per cent of total transistor production in the United States went into communication equipment. As in the United States, the telecommunications equipment market in Europe was not one of the major markets for semiconductor devices.

In addition, the telecommunication end-user market remained mainly an internal market. The major telecommunication equipment producers, such as Siemens, Thomson-Houston, ITT, Philips, Plessey, and GEC, produced semi-conductor components in-house.[63]

The vertical integration between telecommunication production and semi-conductor production was supplemented by a quasi-vertical integration between the telecommunication production of electronics firms and the operations of telephone companies (common carriers). The reason for this quasi-vertical integration can be found in the fact that in Europe the operating telephone companies (common carriers) were public and telecommunications were con-sidered strategically important. The governments of these countries therefore pursued a policy of 'technological sovereignty' in this field. As a result, all of the carriers relied on procurement from firms which had production facilities within the country. Moreover, because contracts were awarded to suppliers according to quotas that were based on existing market shares (OECD, 1981), a stable oligopoly was created in every country; there were few producers and the industry was characterized by a very low rate of entry. In addition, the service providers initiated and subsidized many development efforts of these pro-ducers.[64] As a result, a pattern of quasi-vertical integration developed.[65]

However, the protected in-house markets for semiconductors were never used by telecommunication firms as a pull to innovation. The relatively safe and pre-dictable semiconductor demand was the result of the rather stable oligopoly at the final product stage and of the in-house requirements of semiconductors in these firms. This demand, however, did not stimulate innovation at the com-ponent level because the common carriers that purchased telecommunication equipment from these producers were characterized by a 'conservatism' and pru-dence toward the use of new equipment.[66] As a result of this, these carriers were able to influence the telecommunication equipment producers in a 'con-servative' direction, concerning the introduction of new components into telecommunication equipment.[67]

6 Public policy factors

During the 1950s and early 1960s government support for R&D and production engineering measures (PEM) in Europe was limited. In 1965 (the earliest data available), the British and French governments were the leading supporters of R&D and production of semiconductors in Europe, as Table 4.13 shows.[68]

By contrast, specific policies of R&D support and PEM for the semiconductor industry were totally absent in both Germany and Italy during the 1950s.[69] In Germany, even after the Allies lifted the prohibition against military R&D in 1956, the government preferred to channel its funds into the aircraft industry, while in Italy the government never tried to establish an R&D policy.

Table 4.13 Government funding of the R&D expenditures of firms in semiconductors, by country, 1965 ($ millions)

Country	R&D expenditures of industry	Share of government support (%)
United States	90	na[a]
Great Britain	22[b]	40[c]
France	9	33
West Germany	na	negligible
Italy	1	0
Japan	na	0

Notes: [a]The U.S. Government supported R&D in integrated circuits with $18m.
[b]All active components.
[c]All components.
na = not available.
Source: OECD (1968), p. 157.

In France funds were channelled through the Ministry of Industry, the military and the PTT into Thomson-Houston, COSEM (CSF) and Silec; this support did not produce significant innovations in semiconductors, however. In Britain, support for R&D in active electronic components was channelled through the Inter-Service Committee for the Co-ordination of Valve Development (CVD), a military committee, established in the late 1930s.[70] It has been claimed that 'the overall CVD effort [had] declined in real terms [beginning in] the 1950s though it is probable that this decline [was] less marked in the semiconductor sector' (Golding, 1971, p. 346). Total CVD expenditures reached approximately $5.04 m. in 1963.[71] Generally, the entire cost of projects, plus an agreed profit margin, were paid by CVD, although the profit margin was sometimes exchanged for the proprietary rights of the patents which resulted from the project. Contractors doubtlessly benefited from CVD contracts. However, 'on the device development and production levels . . . government funding in industry appears to have had a very limited impact' (Golding, 1971, p. 355). The recipients of CVD funds included GEC, Mullard (Philips), STC/STL (ITT), Plessey, Ferranti and, later in the 1950s, the English subsidiary of Texas Instruments. In 1957 R&D support for GEC, for example, totalled $210,000, approximately half of its total R&D expenses for that year; in 1962 this figure was increased to $420,000, nearly 50 per cent of its R&D expenses. Plessey, on the other hand, received support from the military for research on silicon in 1953; throughout the 1950s its R&D effort continued to be partially funded by the military.

The two most important government research facilities in Britain were the Royal Radar Establishment(RRE) of the Ministry of Technology and the Service Electronics Research Laboratory (SERL). The RRE was concerned with the application of electronics to military equipment and civil aviation. RRE began working on silicon material soon after Bell announced the discovery of the transistor; yet, although it realized the potential of integrated circuits based on silicon, RRE also began working on thin film and gallium arsenide in the late 1950s. SERL, on the other hand, was responsible for monitoring some of CVD's contracts. Together with RRE, it began to work initially on silicon. Later, however, it switched to germanium. Then, with the development of silicon-diffused transistors, it turned its attention back to silicon, only to switch again to gallium arsenide in 1960.

These government laboratories in Britain had the function of performing research: 'Government laboratories were far less likely to produce anything of commercial value because, at least until recently, this was not considered to be a part of their belief' (Golding, 1971, p. 352). SERL and CVD were convinced that gallium arsenide would become the basic material of transistors and radios; they therefore supported research in this direction. The R&D efforts and the capabilities built while doing research on gallium arsenide would prove useful in the quick diffusion of the Gunn diode in 1963. However, in terms of support for technologies with commercial applications, the choice of focus on gallium arsenide, rather than on silicon, in the late 1950s and early 1960s, was a major policy error.[72]

A comparison of American and British R&D and production refinement policies highlights some important distinctions between American and European public policies.[73] First, the size of American support was much greater than that of either the British or the European case generally, but particularly during the 1950s. Second, the timing of policies was different: while the United States was pushing the missile and space programs in the second half of the 1950s/early 1960s, Britain was gradually retreating from such programs. Third, American policies were more flexible and more responsive than British policies. Finally, research contracts in the United States focused more on development than on research, while in Britain, as well as in the rest of Europe, such contracts focused more on research and proportionately more funds were channelled into government and university laboratories. These last two factors meant that most R&D projects in Britain, as well as in Europe, were not concerned with the commercial application of the results of R&D.

7 The evolution of the American and the Japanese industries

7.1 *The United States: the technological innovator*

During the transistor period, the American industry was highly innovative. Not only did the invention of the transistor occur in the United States, but most of the major innovations of the period were introduced by American firms (see Table 4.14).[74] In the early stages of the American semiconductor industry, the large majority of American semiconductor producers were vertically integrated receiving tube producers (see Tables 4.15 and 4.16).

Table 4.14 Number of major semiconductor innovations by firm,[a] 1950–1960

Bell	4.5
General Electric	1.5
Philco	1.0
Siemens	1.0
Texas Instruments	1.5
Fairchild	1.0

Note: [a]Innovations by two or more producers are shared equally.
Source: Tilton (1971), pp. 16-17.

Table 4.15 United States semiconductor patents awarded to firms in the United States, 1952-1960 (Total = 2,775)

	Share (%)
A. Receiving tube firms	40.5
Bell	15.0
RCA	10.3
General Electric	5.8
Westinghouse	3.8
Sylvania	2.8
Philco-Ford	1.5
Raytheon	1.3
B. Vertically integrated producers	13.4
IBM	2.7
Motorola	1.7
Hughes	2.2
Honeywell	1.4
Sperry Rand	0.9
General Motors	0.7
ITT	1.8
Clevite	0.8
Bendix	0.7
TRW	0.5
C. Merchant producers	
Texas Instruments	14.8
Fairchild	0.03

Source: Tilton (1971), p. 57.

Table 4.16 The market shares of major firms in the United States in semi-conductors, 1957 and 1960

	Percentage of total market	
	1957	1960
A. Vertically integrated receiving tube producers		
General Electric	9	8
RCA	6	7
Raytheon	5	4
Western Electric	5	5
Sylvania	4	3
Philco-Ford	3	6
Westinghouse	2	6
B. Vertically integrated non-receiving tube producers		
Hughes	11	5
Motorola	–	5
C. Merchant producers		
Texas Instruments	20	20
Fairchild	–	5
Transitron	12	9

Source: Tilton (1971), p. 66.

There were some major differences between the evolution of the American and the European industries. These differences were related to supply factors (new, specialized firms), demand factors (public procurement), and public policy factors (R&D support and production refinement contracts).

During the transistor period, several new, specialized merchant producers, such as Texas Instruments and Fairchild, entered the American industry,[75] and exerted a major influence on the evolution of this industry. The routines and strategies of these producers were based on 'electronics' and 'semiconductor' technology, rather than on receiving tube technology or on a condition of dependency on electronics final goods. They therefore had no heritage or tradition stemming from pre-existing electrical or receiving tube technology, or any ties with specific electronics final products. They played an increasingly important role in the development and diffusion of major innovations. In fact, new American merchant producers were responsible for some of the major innovations that occurred during the second half of the 1950s. Texas Instruments, for example, introduced the silicon junction transistor (1954) and the diffused transistor (1956); Fairchild introduced the planar process (1960). Most merchant producers were quick to adopt the metal oxide, diffusion and planar processes.

American public procurement also exerted a profound influence on the evolution of the American industry, both by developing the technological capabilities and focusing the innovative efforts of semiconductor firms, and by fostering the entry of new specialized producers into the semiconductor industry. In fact, 'the presence of a large potential military market increased the rate and influenced the direction of technical change in the 1950s and 1960s.'[76] The Army Signal Corps and the Air Force missile program were the major buyers of semiconductors during this period.[77] The large military markets and, more specifically, the large volume orders awarded to single firms in the United States gave the major contractors incentives for innovation, and, later on, cumulative experience in semiconductor R&D and manufacturing. This allowed them to reduce costs (through the learning curve phenomenon).[78] In addition, because the American military had specific requirements to satisfy through the procurement of semiconductors, it established clear targets for the new devices. As a result, semiconductor producers knew exactly which technological characteristics the military market had. Moreover, because military and civilian technology did not differ extensively during the 1950s and early 1960s, American firms that produced devices for the military/space market had a competitive advantage over European firms in the civilian market as well: both final markets needed smaller, lighter, more reliable and lower power consuming devices (Levin, 1982). The silicon transistor, in fact, was one of the major innovative results of the military and space programs that gave the American industry a strong advantage over the European industry.[79] Moreover, the American military and NASA fostered the entry of new firms into the industry by purchasing products from new firms. The American military, however, did not grant European and American firms the same terms of 'entry'. In many cases, military negotiations did not imply competitive biddings and even when such biddings were required, the 'Buy American' clause (Stoke, 1968) forced foreign firms to bid 6 per cent, and in some cases even 12 per cent, below the lowest bid of American companies. Even the prospect of entering the military market through subcontracting was difficult because many American firms were afraid of possible shortages from European producers. Moreover, NASA awarded contracts to foreigners only if there was an absence of producers in the United States.

The computer and telecommunication markets exerted a less powerful influence on the industry than did public procurement. The demand for semiconductors coming from the American computer industry remained limited during the 1950s. This demand, however, was larger and grew faster than it did in Europe, partly because of the effects of American public policy in support of the industry. Government contracts in the United States, in fact, covered 100 per cent of total computer production in 1954, 60 per cent in 1960 and 47 per cent in 1963.[80] Moreover, American space and defence agencies supported the entry of new firms and aided the survival of several existing computer firms in a market dominated by IBM (Schnee, 1978). Finally, these American agencies funded the R&D of private firms for military purposes; private firms lacked incentive to finance R&D because they did not envision the existence of a commercial market for computers (Katz and Phillips, 1982). The first commercial computers developed in the United States, in fact, were by-products of projects developed for the government. In the case of IBM, space and defence contracts comprised 60 per cent of the firm's total R&D expenditures during the 1950s (Schnee, 1978). These policies had two major effects on the evolution of the American semiconductor industry. First, as the American computer market was enlarged and American producers became the technological and commercial leaders in the industry, the production of diodes and transistors increased. Moreover, by the time transistors began to be used in computers, there was a large, established, American-dominated computer market, ready to absorb the new devices. Second, these policies accelerated the substitution of electron tubes with semiconductors in the second generation of computers.

The American telecommunication market also had a significant influence on the rate and direction of technological change because it was dominated by an innovative firm that was active in semiconductor technology—ATT. ATT was vertically integrated into the R&D and production of semiconductors through Bell-Western Electric. During the 1950s, Bell-Western Electric had an outstanding innovation record in semiconductors. Bell Laboratories invented the transistor, while Western Electric introduced the grown-junction transistor (1951), the oxide-masking and diffusion processes (1955), and the epitaxial transistor (1960).[81]

Conversely, the consumer market did not exert a significant influence on the rate and direction of technological change in the American semiconductor industry, because it was relatively less important in the United States than it was in Europe and because it was increasingly penetrated by Japanese imports. Therefore, the demand for semiconductor devices that came from American firms remained limited.

Finally, American public policy in support of the domestic semiconductor industry was strong, consistent and successful. American public procurement has just been discussed. American R&D support and production refinement contracts, mainly from the military, were directed toward devices with well-defined characteristics: they had to be of a lower weight and smaller size, perform better and be more reliable than previous products. Military support was also highly flexible and responsive to technological developments (Levin, 1982). Between 1955 and 1961, the total amount of funds directed towards R&D and production refinement reached $66.1m. (Tilton, 1971), as Table 4.17 shows. In addition, beginning in the early 1950s, the Army Signal Corporation supported pilot productions at major vertically integrated corporations (Western Electric, General Electric, Raytheon, RCA and Sylvania). They also supported the diffusion of technological information and, later in the 1950s, production refinement in the manufacture of transistors (Kraus, 1973; Golding, 1971; Levin, 1982). Finally, the Air Force supported R&D in silicon transistors for aircraft and missile applications.

Table 4.17 Direct United States government R&D and production capability
support, 1955–1961

	1955	1956	1957	1958	1959	1960	1961
R&D	3.2	4.1	3.8	4.0	6.3	6.8	11.0
Production refinement							
Transistors	2.7	14.0	0.0	1.9	1.0	0.0	1.7
Diodes and rectifiers	2.2	0.8	0.5	0.2	0.0	1.1	0.8
Total	8.1	18.9	4.3	6.1	7.3	7.9	13.5

Note: The American government supported R&D in integrated circuit, with an average
of $18m. in the period 1960–65 (OECD, 1968, pp. 57–8).

Source: US Department of Commerce; Business and Defense Services Administration;
Semiconductors: US Production and Trade 1961.

7.2 *Japan: success in the consumer market*

During the transistor period, the Japanese industry did not play a major role in
the international semiconductor industry. The only Japanese firm to introduce
a relevant innovation was Sony, with its tunnel diode.

The structure of the Japanese semiconductor industry was similar to that of
the European industry: it was composed mainly of a few, vertically integrated
producers, such as Hitachi, Toshiba, Matsushita, Nippon Electric and Mitsubishi.
As with the European and American industries, these vertically integrated producers were the first entrants into the semiconductor industry in the mid-1950s,
and they dominated the industry during the late 1950s. Firms that did not
produce receiving tubes were comparatively few: Sony, Fujitsu, Sanyo, Japan
Radio and OKI (vertically integrated producers) and Kyodo (merchant producer). Contrary to the European industry, however, the Japanese industry
was not at the technological frontier in electronics at the time the transistor
was invented.

Not only the structure of the industry, but also the structure of demand
in Japan was similar to that in Europe. The consumer market was the major
market for semiconductors. It grew rapidly, owing to the successful performance
of Japanese exporters of electronics consumer goods. During the 1950s and
1960s, Japan exported large numbers of transistor radios using transistor devices
(Tilton, 1971) and later, black and white television receivers (Levy, 1981) to
the United States. Europe was partially isolated by the export flows of Japanese
television receivers, because European countries used transmission standards
that were incompatible with the System M used in the United States and Japan.

The Japanese semiconductor industry was not the direct target of specific
government policies; it was supported, however, by the general policies of the
Japanese government.[82] The government, in fact, protected the Japanese market

from the entry of foreign subsidiaries, and favoured the purchase of American licences by Japanese producers. Even in the case of joint ventures, Japanese firms had to hold over 50 per cent of the capital.[83] As a result, Japanese firms remained the only producers located in Japan (Tilton, 1971). They were therefore able to profit from the low cost of labour and from direct access to the Japanese market.

Conclusions

The birth of the semiconductor industry emphasizes the fact that a new industry is not born out of nothing, but out of an existing industrial structure. At a general level, in fact, the structure of the existing industry may strongly influence the structure and the evolution of the new industry. In the case of the semiconductor industry, an established oligopolistic industry (the electrical industry producing receiving tubes) originated the new industry and perpetuated the former oligopolistic structure during the early stages of this industry. These initial structural conditions characterized not only the European case, but also the American and Japanese cases.

In the case of the semiconductor industry, the continuity of the previous industrial structure derived from the fact that receiving tube producers had advanced technological capabilities in electrical and electronic technologies, were convinced of the future importance of the new technology, and realized that these new products addressed the same market as their former products. Therefore, contrary to what has been predicted by authors such as Schumpeter, Abernathy and Utterback, about the presence of several new small firms in the early stages of a new industry, the early stages of the semiconductor industry were characterized by the presence of a few large, established vertically integrated firms.

This initial oligopolistic structure continued to characterize the world semiconductor industry until the second half of the 1950s. Established receiving tube producers continued to be the major innovators in the industry. These producers were electronics final goods producers that had integrated into semiconductor production because of tradition, the availability of redundant capability and the desire to have a secure source of component supply. Although American firms were more innovative than European firms, the latter were not at any technological disadvantage relative to their American counterparts during this period. Many European, American and Japanese corporations, in fact, had similar areas of specialization (e.g. Siemens/General Electric/Westinghouse, Philips/RCA/ Matsushita), followed similar strategies, and obtained similar economic results.

During the second half of the 1950s, however, demand and public policy factors affected the dynamics of the European and American industries differently. The differences were due to the importance of the consumer market in Europe (and Japan) and to public policy in the United States.

During the late 1950s, established vertically integrated receiving tube producers continued to dominate the European industry and to focus their production on the consumer and, to a lesser extent, industrial markets. As a result,
they were successful in the production of germanium alloy transistors and silicon
controlled rectifiers. They also innovated incrementally in post-alloy diffused
transistors and in mesa transistors.

By contrast, during the mid- and late-1950s, major changes occurred in the
American industry. The American military created two new markets for semiconductor devices: one directly (the public procurement of devices), and one
indirectly (the computer industry). These new markets added large, and partially protected markets to established electronics final markets such as the
consumer, telecommunication and industrial markets. Public procurement
stimulated the introduction of new technological alternatives, such as the use of
silicon in transistors and the fabrication of a circuit on a single semiconductor
device. Both markets permitted the entry of new merchant producers, such as
Texas Instruments. These firms were not bound to any established technology
or market and were therefore able to commit resources and focus their technological and productive efforts on the requirements of these new markets.

Notes

1. See Nelson (1962) and Braun and MacDonald (1982).
2. The Italian industry was, and remained, the weakest of all the major
   European electrical industries. Until the Second World War the Italian
   electrical industry was highly dependent on the production of the Swiss
   firm Brown Boveri. After the war a domestic industry arose. The dimensions of Italian producers, however, remained small compared to those of
   their international competitors. The major producers included: CGE, Ercole
   Marelli, Franco Tosi, and two public enterprises, Ansaldo and Breda.
3. Mainly Schockley and, on a more general level, Kelly. See Nelson (1962).
4. One of the inventors of the transistor has claimed that Bell's management
   even impaired innovations in several cases (Braun and MacDonald, 1982).
5. The point-contact transistor was 'essentially a V-shaped wire in a germanium
   crystal' (Golding, 1971, p. 69). See also Nelson (1962).
6. It was developed at Bell and produced by Western Electric in 1951 (Tilton
   1971). The material used in the device changed from polycrystalline to
   single crystal while it was slowly withdrawn from the melt. These transistors
   produced less noise.
7. It was developed by Texas Instruments in 1954.
8. The silicon-based grown-junction transistor was of great interest to the military market.
9. The alloy process and the alloy-junction transistor were developed by
   General Electric and RCA in 1952 (Tilton, 1971). Dots of indium were
   alloyed to the opposite sides of a wafer of germanium. The jet-etching

process and the surface barrier transistor were developed for high frequencies. They were developed by Philco in 1954. The germanium wafer was eroded by an electro-mechanical method. Because of the thinness of the base layer, the surface barrier transistor was more delicate and more expensive than the alloy-junction transistor. However, it could reach the highest frequency that was commercially available around 1955 (Golding, 1971).

10. The oxide-masking and diffusion processes were developed by Bell and General Electric in 1955–6. Bell organized a symposium in 1956 on these processes. In the oxide-masking and diffusion processes an impurity was diffused from the vapor phase on the semiconductor. Regulation of the time and temperature of the furnace controlled the penetration of the impurity. Advanced photographic techniques and a mask of silicon dioxide determined the areas on the semiconductor which were open to impurities. See Braun and MacDonald (1982).

11. In the drift transistor the impurity concentration was diffused smoothly across the base region. This transistor also had a good high-frequency response (Young, 1979).

12. The transistor's base was diffused into a substrate, then the portions around the base were etched away so that a plateau was left above the substrate. The emitter was diffused inside the base region. The mesa transistor was relatively cheap to produce because batch production was possible in all stages of its production, except the electrical connections stage (these connections had to be made manually). There was also potential for the contamination of the mesa surface. The mesa transistor handled high-frequency ranges and dissipated heat readily (Young, 1979).

13. Abernathy (1978) analyses this phenomenon for the automobilie industry. In addition, see Hill and Utterback (1979).

14. Although Schockley had produced a theory about junction transistors. Nelson (1962) and Braun and MacDonald (1982).

15. See Webbink (1977). The shipments of transistors declined from 193 million units (1960) to 131.8 million units (1961) and of diodes and rectifiers from 272.3 million units (1960) to 178.8 million units (1961). In value, transistors passed from $316.2m. in 1960 to $313.9m. in 1961 to $303.5m. in 1962, while diodes and rectifiers, after an increase in value to $210m. in 1961, decreased to $184.5m. in 1962 (pp. 7, 11).

16. Electronics Industry Association (1979), pp. 106–7.

17. Apparent consumption is equal to production plus imports minus exports.

18. Except for a few years during the second half of the 1950s.

19. Except for a few years during the second half of the 1950s.

20. Finan (1975) confirms this statement by showing that only 11 per cent of the total European consumption of semiconductors in 1960 was supplied by exports from the United States.

21. Table 4.4 provides a list of the main entrants in the European industry by country. Tilton (1971) provides a complete list of entrants by year of entry.

22. Siemens had already been producing photo diodes and germanium diodes.

23. In 1953–4, Siemens built a factory in Munich for the production of electron

tubes, semiconductors and passive devices. With the addition of this Munich branch, Siemens had four major productive centres: Heidenheim, Munich, Berlin and Regensburg. Sources: interviews and company reports.
24. Freeman (1965), Golding (1971), Sciberras (1977) and interviews have provided the material for this section on Great Britain.
25. In the late 1950s Plessey's in-house use of its semiconductor production reached approximately 10 per cent. In 1958, Plessey obtained a license for producing thyristors from General Instruments.
26. The categorization of SGS as a vertically integrated producer is ambiguous; SGS was created through a joint venture between Olivetti and Telettra (FIAT). Telettra was an electrical firm which produced telecommunication equipment. However, its use of SGS' components remained limited. Olivetti, on the other hand, was an office equipment producer which used mechanical technologies and which was reluctant to switch to electromechanical and electronic technology. Source: interviews.
27. Theoretically, new semiconductor production processes and devices could be protected by patents. During the 1950s and early 1960s, however, semiconductors were still discrete devices and could be described precisely in patent claims. For a detailed discussion of patents in the semiconductor industry, see Levin (1982) and MacDonald (1981).
28. These similarities greatly reduced the cost of the transfer.
29. For a discussion of British firms, see Golding (1971).
30. Philips was a consumer electronics firm like RCA; Siemens was a telecommunications equipment producer like Bell-Western Electric; AEG-Telefunken was a consumer electronics firm like General Electric, and so on.
31. For an overview of the factors which may induce a producer to make direct foreign investments rather than to export, see Rugman (1981).
32. Source: interviews. The following chapter examines the way in which these technological beliefs were challenged by a new process stemming from the diffusion process: the planar process.
33. Source: interviews. The manufacturing process for this transistor evolved from the diffusion process; two pellets of lead mixed with antimony were placed on a crystal of p-germanium. Then, a small quantity of aluminium was applied to the right pellet. The process of diffusion and alloy which followed had to take place at very high temperatures (780°). The right pellet formed a p-n junction and served as the emitter while the left pellet formed the base. Finally, the n-germanium outside the junction was etched away and terminal wires were soldered onto the pellets. Source: Philips.
34. Source: interviews. This situation is very similar to that of some American producers, like General Electric and Westinghouse.
35. This section is based on Golding (1971) and interviews.
36. Golding (1971). The move of International Rectifier US was caused by increased competition in heavy power equipment applications in the United States by General Electric and Westinghouse. See Finan (1975).
37. Most of the information in this section comes from interviews with the managers of European corporations, experts, public officials and scholars. The OECD confirms this view: 'In Europe, the decision to enter a new

field like semiconductors . . . resulted . . . from the firm's realization that these new technologies were the important ones, and that they should be pushed rapidly. In other words, the question was essentially one of judgement, both business-wise and technology-wise' (OECD, 1968, p. 97).

38. As will be shown later, a similar type of conjecture induced Siemens to begin the production of computers.

39. During the 1950s, the average share of semiconductors purchased from external sources by these corporations was lower than the share of semiconductors sold to external customers. Source: interviews.

40. Profitability in the production of receiving tubes was assured by a stable oligopoly around the world. The production of small signal diodes (and, in particular, germanium diodes) rapidly became standard. The technology of the device was simple and technological change was slow.

41. Theoretically, this knowledge could have been traded between the electron tube producer and the new entrant. However, the economic and technological evaluation of this knowledge presented serious difficulties and therefore such exchanges would have involved high transaction costs.

42. The following analysis is based on interviews with the managers of several European corporations, experts, public officials and scholars.

43. Several economists have analysed the incentives of firms to integrate vertically under such conditions, i.e., in the presence of small input markets. Stigler, one of the main authors in this group (Stigler, 1951), claims that when the size of the market for inputs is small, a specialized input producer may not survive; this holds true even if the activity has increasing returns. As a result, a firm must integrate vertically and produce its inputs in-house. If the size of the market is large, the increasing returns activity can be profitably performed by a specialized firm. Two major assumptions lie at the base of this type of analysis: market situations must be compared in a static rather than in a dynamic framework and transaction costs must be absent. Adelman (1955) provides a different interpretation; he offers a dynamic view of the integration of firms during the evolution of an industry. Adelman claims that when both the inputs and the final goods industries are small and new, firms will specialize, and when the final goods industry grows, final producers will integrate upstream. This interpretation is more apt to explain the situation of the semiconductor industry in its early stages.

44. See Arrow (1975). These incentives are even stronger when information about the future supply of inputs, needed by producers some time in advance of the actual supply, is controlled by the input producer.

45. Much of the information in this section comes from interviews with the managers of various European firms.

46. Other aspects of the internal organization of vertically integrated corporations, not discussed here, should be noted: for instance, the specific type of R&D and productive organization. GEC and AEI provide two illustrations of this point. Wembley (GEC) was the major centre in Britain for R&D in semiconductors. However, the Wembley offices were separated from the location of semiconductor manufacture; this situation created communication problems. In addition, R&D projects in semiconductors were financed

by GEC's final product groups. Similarly, the production of semiconductors in AEI remained divided into several equipment groups. Such situations created rivalries between the various activities within these firms and made the semiconductor division dependent on the decisions taken by other groups. See Golding (1971).

47. As will be shown later, the structure of demand in Europe was more similar to that in Japan than to that in the United States. In the United States, in fact, public procurement and the computer market were relatively more important.

48. In 1957 Siemens Electrogaerate was opened; it produced domestic appliances, radios and TV sets and it used a minor part of Siemens' semiconductor production. Siemens Electrogaerate remained of secondary importance within Siemens, however. (Source: interviews.)

49. There was an important difference between the American and European markets. In Europe the tuner in a radio receiver was used as a pre-amplifier. In the United States the tuner was constituted by a diode and the amplification function was performed separately. So, in the United States there was no demand for high-frequency pre-amplification devices. Source: interviews.

50. As will be shown later, the situation was quite different in the United States.

51. See Table 4.9. More conservative estimates of the share of public procurement during the late 1950s have established an approximate value of 15–20 per cent for Britain (Golding, 1971, pp. 141–2) and 39–48 per cent for the United States (US Department of Commerce, 1960, in Levin (1982), p. 60).

52. See Keck (1976). The Japanese case is similar to the German.

53. In the United States for example, the work of Aitken on the Mark I dates back to 1937. This work was followed in 1945 by the construction of the ENIAC, a computer which was later given to the Army, and in 1951 by the development of the UNIVAC I, the first commercial computer. The UNIVAC I was delivered by the UNIVAC division of Remington Rand (Brock, 1975). During the 1940s and 1950s work on computers within Europe was at an advanced stage. In Germany, K. Zuse built the Z3 computer in 1941, five years before the ENIAC was built in the United States. In 1950, work on computers began in the universities of Darmstadt, Göttingen, Munich and Dresden; the first workable model, Göttingen University's G-1, was completed in 1952 (OECD, 1969). In Britain the William tube, which led to memory systems based on cathode tubes, was invented in 1946 and in 1948 the University of Manchester built its first computer. Following this discovery, Ferranti and the University of Manchester agreed to co-operate on the construction of a commercial computer and in 1951 Ferranti delivered a computer, the Mark I, to the University. In addition, Lyon's Tea Shops built the first computer designed for business, the Leo, in 1951 (OECD, 1977). Finally, in France, after the pioneering work of Coffignal in 1951, Bull produced the Gamma 3 computer which was the first computer in the world to utilize germanium diodes (OECD, 1977).

54. As a top Siemens manager claimed, 'in the near future data processing installations will be as much an integral part of power engineering systems as motors, transformers or measuring instruments are today.' (Von Weiher and Goetzeler, 1984.)

55. Siemens decided to enter the electronic data processing market in 1954. In 1957 the first trial runs of the 2002 Computer, a fully transistorized data processing system, were made, and in 1959 one of the 2002 systems was installed at the Aachen Technical University and one within Siemens itself. A special type of germanium alloy-junction transistor was produced by Siemens and used in the 2002. The 2002 was followed by the 3003 computer. Neither of these lines (the 2002 or the 3003) was commercially very successful, however. Von Weiher-Goetzeler (1984); Siemens Berichte (various years); interviews.

56. In the late 1950s, Telefunken received a licence from RCA to produce the TR4, a large specialized computer, using the germanium drift transistor which had already been manufactured by Telefunken. The TR4, however, was not commercially successful either. In addition, during the 1950s, Standard Electric was forced to discontinue its production of the Zebra computer after only a short period (OECD, 1969).

57. During the development of Bull's Gamma 60, the transistor generation of computers based on transistors appeared on the market. Bull was caught unprepared for the passage from receiving tube to transistor computers. It decided, however, to delay the introduction of its new line (Gamma 60) in order to convert it to the use of transistors. This delay meant that the Gamma 60 was introduced only one year before IBM successfully launched the 7000 and the 1400 series; thus, the market for the Gamma 60 was severely reduced soon after its introduction (OECD, 1977 and Hu 1973).

58. The situation with regard to the British computer industry was similar to that of the British semiconductor industry: a large number of firms (greater than the European average) began the commercial production of computers during the 1950s. After the attempts of BMT and LEO to begin computer production in the 1950s, Plessey and Ferranti emerged as the major domestic producers of computers. Plessey and ICT obtained licences from American producers, while Ferranti developed an internal technological capabilty (OECD, 1977). Ferranti's Argus computer for process control resulted from a government contract to produce a missile control computer (Freeman, 1965). Ferranti, however, exited from the computer business in 1963, selling its activity to ICT.

59. Olivetti produced a receiving tube computer, the ELEA 9003, in 1960. This was followed in 1961 by the development of the ELEA 6001, a computer for scientific use, and by the construction of a new factory for computer production in Pregnana. By 1964, Olivetti controlled approximately 25 per cent of the Italian computer market. The Italian market, however, remained small and economies of scale could not be attained. Moreover, Olivetti lacked funds for the support of a large-scale R&D and manufacturing effort. In fact, the purchase and subsequent restructuring of the office equipment producer Underwood in late 1959 drained Olivetti's

funds away from the financing of other activites (Soria, 1979). The financial troubles created by this situation were complicated by the underestimate on the part of management of the future impact of electronic computer technologies on office equipment production, and led to the sale of 75 per cent of the Olivetti electronics division to General Electric in 1964 (*Business Week*, 26 October 1974).

60. One of the few exceptions to this pattern was the relationship between IBM and Philips: IBM purchased semiconductors from Philips throughout the 1960s.

61. During the 1950s some American computer manufacturers were also vertically integrated and produced part of their semiconductor requirements in-house. This is true, for example, of RCA and GE.

62. Olivetti, for example, used part of SGS's production of alloy transistors in its ELEA computers. However, Olivetti never treated SGS as a dependent captive supplier; it purchased semiconductors from other firms as well (e.g. Fairchild).

63. Source: interviews. As will be shown later, the Japanese case is very similar to the European case in this respect.

64. In Britain, the Post Office, working through the Bulk Supply Arrangement, awarded contracts to GEC, Plessey and STC (ITT) (Diodati, 1980). In West Germany, the Deutsche Bundespost was supplied by Siemens and SEL (ITT). In France, the PTT purchased equipment from two supply cartels: Socotel for switching equipment and Sotelec for transmission. Finally, in Italy there were several public buyers and the quotas were divided geographically among five producers: SIT-Siemens (state-owned), Telettra (Fiat), FATME (Ericsson), Face Standard (ITT) and GTE (Lizzeri and De Brabant, 1979). Only four multinational corporations manufactured telecommunications equipment in Europe: Philips, Ericsson, ITT and GTE.

65. Quasi-vertical integration refers to a situation in which there is a 'close and continuous contact between service providers and equipment manufacturers in the innovation process' (OECD, 1981, p. 140; and Diodati).

66. Source: interviews.

67. As will be shown later, a different situation characterized the American telecommunication market.

68. Of course, the relative positions of the United States and Europe, based on the level of support for R&D and production in 1965, was not the same as their relative positions during the 1950s. During those years, American involvement in the support of the industry remained high, while European involvement was almost non-existent.

69. This is also true for Japan.

70. Golding (1971); Freeman (1965, 1982). The British case is well documented in a series of studies. A close examination of British policy during the 1950s and early 1960s and a comparison of British and American policy during this period may provide useful insights into the differences in the evolution of the European and American semiconductor industries. The following section is based on Golding (1971) and interviews.

71. This figure includes expenditures for other active devices, such as electron tubes (Golding 1971).
72. Golding (1971). It is interesting to note that, at the same time, the American military was pushing for silicon devices.
73. American public policy is examined in more detail in the next section.
74. For a list, see Tilton (1971).
75. For a list, see Tilton (1971).
76. Levin (1982), p. 58. See also Braun and MacDonald (1982), Asher-Strom (1977), Utterback and Murray (1977).
77. The emphasis of procurement in the United States changed from one based on aircraft (36 per cent of total Department of Defense procurement) in 1955, to one based on aircraft and missiles (22.5 and 23.4 per cent respectively) in 1960. Source: OECD (1968), p. 57.
78. The government market accounted for 52 per cent of Texas Instruments' semiconductor sales between 1951 and 1959 (Freeman, 1965) and 60 per cent of Transitron sales in 1960 (Golding, 1971, p. 169).
79. The silicon transistor was initially aimed at military equipment that needed transistors that were capable of performing at high temperatures (Levin, 1982; Mowery, 1983).
80. During the second half of the 1950s, government contracts accounted for 30 per cent of IBM's sales (Schnee, 1978).
81. Tilton (1971). For a discussion, see also Levin (1982).
82. In 1957, however, Public Law 117 (the Electronics Industry Provisional Development Act) reinforced MITI's leading role in the policies for the electronics industry. In addition, a few years earlier, a transistor computer logic research project had been co-ordinated by MITI (SIA, 1983).
83. For example, in Matsushita (70%)-Philips (30%), Japan Radio (67%)-Raytheon (33%), International Rectifier Corporation of Japan (61%)-International Rectifier USA (39%).

5   The integrated circuit period: the decline of the European industry

In this chapter the evolution of the European semiconductor industry during the integrated circuit period will be analysed. The integrated circuit period covers the years between the introduction of the planar process and the integrated circuit (1959-61) and the introduction of the microprocessor (1971).

The chapter is organized in the following way: a discussion of the technological regime, the major innovations and the technological trajectories is contained in Section 1. The history of the evolution of the European semiconductor industry and of its decline is presented in Section 2. The major factors affecting the evolution of the European industry (supply, demand and public policy) are analysed in Sections 3, 4 and 5. Finally, the evolution of the American and the Japanese industries is discussed in Section 6, for purposes of comparison with the European industry.

1 Technological regime, major innovations and technological trajectories

The integrated circuit period begins with the introduction of two radical innovations: the planar process and the integrated circuit. The first radically altered the manufacturing process of semiconductors. The second introduced a completely new product to the industry.

The planar process allowed the batch process, and therefore the large-scale production, of semiconductor devices. It was developed by Noyce at Fairchild in 1960.[1] This process was based on the older oxide-masking and diffusion processes. The planar process had several advantages over the old mesa technology. First, its oxide mask was protected against contamination by impurities. Second, the protection of the junction was improved with the addition of the oxide layer. Third, electrical connections were not carried out by hand but by depositing an evaporated metal film: as a consequence, the planar transistor could be produced by batch technique, transistors could be produced on a large scale, production costs were lowered and the reliability of the transistor was improved. The planar process became the standard means used by semiconductor producers to manufacture integrated circuits. However, it had two limitations. First, it could only be applied to silicon material because germanium was not compatible with the oxide layer that was grown on the surface of the device: as a result, silicon, rather than germanium, became the major material used in the industry. Second, it was not applicable to high-power devices which used diffusion and epitaxy.

The integrated circuit consisted of a whole circuit placed on a single silicon wafer. The germanium integrated circuit was introduced by Kilby at Texas Instruments and patented in 1959, while the silicon-planar integrated circuit was introduced by Fairchild and Texas Instruments in 1961. Initially, integrated circuits had been made of germanium. Later, however, the planar process began to be used extensively and integrated circuits were manufactured with silicon material.

The technological regime of this period was based on the use of silicon as material, on the planar as manufacturing process, on the integrated circuit as product and on the integration of components and system as product concept. During the 1960s, silicon replaced germanium in most semiconductor devices, the planar substituted the oxide-masking and diffusion processes and the integrated circuit took the places of several diodes and transistors connected in a circuit.

The oxide-masking, diffusion and planar processes provided a starting point for the creation and refinement of an efficient batch process technology and for the introduction of a number of new innovations. Two major groups of products, in fact, were available during the 1960s: discrete devices and integrated circuit devices.[2] The existence of alternative technologies and alternative product types, however, meant that only those firms having large R&D capabilities and production facilities could produce the entire range of products (discrete devices, linear and digital integrated circuits), and eventually reap economies of scope in production.

The increasing integration is related to the increasing complexity of semiconductor devices. With the introduction of the integrated circuit, a semiconductor device could now perform functions which, during the transistor period, had been performed by circuitry. During the integrated circuit period the elements of this circuitry were grouped and placed on a single, monolithic semiconductor chip and circuit design was carried out directly on the semiconductor device. This increase in integration meant that a steadily growing number of both circuits and functions could be contained on a single semiconductor device.

As a consequence, during the 1960s, custom semiconductor devices became increasingly important. The increasing integration of semiconductor devices, in fact, meant that these devices could not only perform an increasing number of functions, but also that they could be custom-designed to the specific needs of a few customers. The distinction between standard and custom devices was closely connected to the distinction between digital and linear integrated circuits. Digital integrated circuits consisted of a large number of identical components; their most common components were transistors and diodes and, therefore, the standardization of the device was easy to attain. Linear integrated circuits, on the other hand, were more difficult to standardize. These circuits contained a large number of passive components (resistors, inductors and capacitors) which were more difficult to manufacture and which occupied more space than active devices.

During the integrated circuit period, the advances that were made in technology, at both the process and product levels, required more applied R&D efforts, more engineering and design capabilities and less basic R&D effort and scientific know-how (Braun and MacDonald, 1982; Levin, 1982). Moreover, the complexity of semiconductor processes and devices increased as a result of the possibility of cramming more functions onto a single chip. The development of semiconductor techniques was handled less and less by single individuals and more and more by large organizations.

The major innovations of the integrated circuit period at the product level were centred on the integrated circuit. Digital integrated circuits were the first types of integrated circuits to be introduced. Digital bipolar integrated circuits were the first digital integrated circuits to be produced.[3] In 1961, Texas Instruments introduced the Series 51, a Resistor-Capacitor-Transistor Logic (RCTL),[4] while Fairchild introduced the Direct-Coupled-Transistor Logic (DCTL). One year later, Signetics introduced the Utilogic, Diode-Transistor Logic (DTL)[5] and Westinghouse followed (in 1963) with the 200 Series DTL. The Diode-Transistor Logic had widespread use, especially in the computer industry.[6] In 1962/63 TRW and Motorola introduced the Emitter Coupled Logic (ECL).[7] In 1964 both Sylvania and Transitron introduced a transistor-transistor-logic (TTL).[8] This TTL was an integrated circuit composed of active components; it became the most widely-used digital bipolar integrated circuit. Finally, in 1969, Texas Instruments introduced the Schottky TTL in which a Schottky diode was added to a TTL integrated circuit so that the speed of the circuit could be increased. Digital bipolar integrated circuits were followed by the introduction of digital MOS integrated circuits in the mid-1960s by General Microelectronics and General Instruments.[9]

Linear integrated circuits were produced later than digital integrated circuits. The first linear integrated circuits to be produced were operational amplifiers; they were used in analog computers and were manufactured by Texas Instruments and Westinghouse. Linear integrated circuits were also used in radios, lower-power stereo amplifiers, TV sets and automobiles. They generally used bipolar technology although they could use MOS technology in order to increase the precision and stability of the circuit. Linear integrated circuits were introduced later than digital integrated circuits for four reasons. First, they were dependent on resistors and capacitors, both of which were difficult to integrate. Second, they needed inductors for tuning and filtering, neither of which could be added unless resistor and capacitor networks and additional circuitry were used. Third, they required the interaction of the operating components and the semiconductor substrate; this interaction was more difficult to control in linear integrated circuits. Fourth, the design and manufacturing processes of linear integrated circuits were more difficult than those of digital integrated circuits.[10]

The major innovations of the integrated circuit period at the process level included epitaxy,[11] resistive metal deposition, ultrasonic bonding and encapsulation, annealing for stress relief and the gettering process, isolation, nitride chemical and glass chemical vapour deposition, ion implantation, and the Schottky junction.[12]

The technological trajectory of the IC period was toward miniaturization and integration. Given the fact that an integrated circuit contained some characteristics of a system, the R&D efforts of firms were devoted to increasing the level of integration of a device, and to reducing its size.

This technological trajectory was enhanced by the fact that, with the introduction of oxide-masking, diffusion and planar processes in the late 1950s, technological change in semiconductors became increasingly cumulative.[13] This meant that new technological developments were increasingly based on advances that had been made in the past; the planar process, for example, was based on the masking and diffusion processes and the integrated circuit was based on the planar process. It also meant that firms that had some advantage in either a product or a process range in one period had a high probability of retaining that advantage in the following period.

Contrary to what happened during the transistor period, during the integrated circuit period innovations at the product level became more independent from innovations at the process level. The planar process, in fact, established a well-defined production process, which could be improved but not drastically altered. This process is at the base of a large number of product innovations.

The developments in technology affected semiconductor firms in several ways. First, as semiconductor technology became more complex, organized R&D became more important and required larger amounts of investment and advanced engineering capabilities.[14] Second, because manufacturing and marketing aspects had an increasing influence on innovation in semiconductors, firms began to locate their R&D laboratories close to their production facilities. This move permitted the continual exchange of both information and personnel. It also meant that the bulk of advances made in semiconductor technology were made within firms, rather than within research laboratories. Most of these advances were based on engineering skills and were focused on 'techniques' (the 'proprietary') rather than on 'logy' (the 'public' aspect of technology).[15]

2 The evolution of the European industry: a case of decline

2.1 *The world industry*

During the integrated circuit period, the United States continued to be both the largest producer and the largest consumer of semiconductors in the world. In 1969 55 per cent of world semiconductor production was supplied by US-based companies (Braun and MacDonald, 1982, p. 121). The European countries ranked below the United States and, by the beginning of the 1970s, even below Japan (see Tables 5.1–5.3).

Similar to the transistor period, during the 1960s the diffusion of integrated circuit consumption was much more rapid in the United States than in either Europe or Japan. In 1967, for example, the ratio of the consumption of integrated circuits to the consumption of semiconductors was approximately 1:3 in the United States, 1:20 in Europe and 1:40 in Japan. (See also Table 5.4.)[16]

However, at the beginning of the 1970s, the production and consumption

Table 5.1 Consumption of semiconductors by major world areas (percentages of the world market)

	1960 %	1965 %	1970 %
United States	76	66	53
Western Europe	12	18	22
Asia	10	14	22
Other	2	2	3
Total	100	100	100
Market Size:	$750m.	$1,700m.	$3,000m.

Source: US Department of Commerce (1979) from Finan (1975).

Table 5.2 Consumption of semiconductors by country ($ millions)

	1960	1965	1970	1972
Great Britain	28	72		210
France	27	67	420	114
West Germany	25	52		218
3 Countries	80	191	420	542
Japan	54	132	420	742
United States	560	1064	1547	1708

Source: Finan (1975).

Table 5.3 Production of semiconductors by country—estimates ($ millions)

	1961	1964	1965
United States	607	635	927[a] / 85[b]
Japan	78	139	140
Great Britain	35	66	na
France	32	52	59
West Germany	30	61	47

Notes: [a]Discrete semiconductors
[b]Integrated circuits
Sources: Freeman (1965), p. 89 for 1961 and 1964; and OECD (1968), p. 15 for 1965.

of integrated circuits was still inferior to that of discrete semiconductors in all countries. In 1970 in the United States, for example, the sales of integrated circuits products by American firms totalled $524m. vs. $769m. of sales of discrete semiconductors (Electronics Industry Association, 1979). Digital integrated circuits and bipolar integrated circuits were the most widely-used types of integrated circuits during this period (as Table 5.4 shows).

Table 5.4 Annual manufacturer sales of integrated circuits in the United States —consumption of integrated circuits in Great Britain, 1964–1968

	Annual manufacturer sales ($ millions) United States	Consumption ($ millions) Great Britain
1964	51	1.5
1965	94	3
1966	173	10
1967	273	17
1968	367	26

Sources: Electronics Industry Association for the United States and Golding (1971) for Great Britain.

Table 5.5 Integrated circuits (as percentage of total integrated circuit market)

	1965 %	1970 %
Linear	18	18
Digital	82	82
Bipolar	82	66
MOS	0	16

Source: Mission pour les circuits intègres in Truel (1981), p. A.13.

During the 1960s the prices of transistors and integrated circuits continued to decline in the United States. The average value of a silicon transistor decreased from $11.27 (1960) to $.38 (1970) (Electronics Industry Association, 1979) and the average price of an integrated circuit went from $50 (1962) to $2.33 (1968) (Tilton, 1971, p. 91). This reduction in price was the result of a continuous increase in yield throughout the various stages of the production process.

Semiconductor 'recessions' occurred in 1966–7 and in 1970–71. In the first recession, sales of discrete semiconductors in the American market decreased by 14 per cent; only in 1969 did they begin to increase again. In the second recession, a reduction in the overall sales of semiconductors occurred simultaneously with a major competitive struggle for supremacy in the digital integrated circuit market, 'the bipolar logic war'. The semiconductor recession of

1970-71 affected almost all industrialized countries. In the United States, for example, sales of discrete semiconductors dropped 19 per cent in 1971 and sales of integrated circuits remained almost constant between 1970 and 1971. In Germany sales of discrete devices and integrated circuits dropped 40 per cent and 28 per cent respectively and the production of discrete devices and semiconductors decreased by 57 per cent and 50 per cent respectively in 1971 (Scholz, 1974, p. 130); in France both the sales of discrete semiconductors and the sales of integrated circuits decreased by 17 per cent and 20 per cent respectively (*Usine Nouvelle*, 1976-4 and France-Commissariat Général du Plan, 1976).

The performance of the various countries in international trade in semiconductors differed greatly during this period. The United States and Japan (for most of the 1960s) had trade surpluses, while West Germany, France, Britain and Italy had trade deficits and became less and less self-sufficient[17] (see Figure 5.1 and Table 5.6). International trade in semiconductors was closely linked to the supply of semiconductors by American firms located in Europe. American subsidiaries in Europe increased in number during the 1960s and began directly supplying the European market (as Table 5.7 shows).

Table 5.6 Trade balance in semiconductors by country, 1966 ($ millions)

United States	+66.6
Japan	+10.3
Great Britain	−26.3
France	−17.6
West Germany	na
Italy	na

Note: The trade balance with other countries of the United States in semiconductors in 1966 was +$13.1m. with Great Britain, +$10.5m. with France, +$5.6m. with West Germany and −$0.2m. with Japan (OECD, 1968).
na = not available
Source: OECD (1968), p. 33.

In the following sections the declining trade performance of the various European countries and the evolution of the European semiconductor industry during the 1960s are analysed. The evolution of the European industry during the integrated circuit period is divided into three phases: the reaction of the European industry to the new technology (first half of the 1960s); the penetration of American producers in the European market (mid- to late-1960s); and the European defeat in the digital integrated circuit war (early 1970s).

2.2 The reaction of European firms to the new technology

In the early 1960s, as previously examined, a new technology based on silicon as material, on the planar process as fabrication method and on the increasing

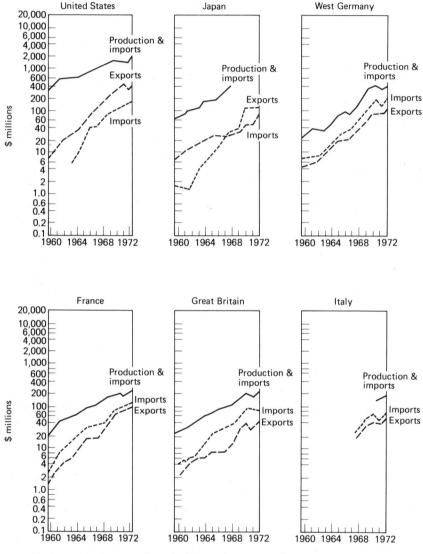

Figure 5.1 Consumption, exports and imports of semiconductors by country, 1960–72. *Source*: Tilton (1971); Dosi (1981).

Table 5.7 Direct and indirect imports of semiconductors of major countries from the United States, 1960 and 1969

	Great Britain-France-Germany-Japan imports from the United States as percentage of the domestic market and number of American factories in Europe			
	Direct imports from US (%)	Direct & indirect imports from US (%)	Cumulative assembling factories	Number of fabricating factories
1960	11	na	4	4
1969	37	40	24	15

Source: Finan (1975) in Braun and MacDonald (1982).
na = not available

integration of devices became dominant within the semiconductor industry. This new technology displaced the old technology that had been based on germanium as material, the alloy-and-mesa processes as fabrication methods and the component conception of the semiconductor device.

Many established firms adapted slowly to the new silicon-planar-integrated circuit technology. This new technology did not seem to represent a great departure from the past. While the transistor had represented a great departure from receiving tubes in terms of material, performance and application (although it was always considered a typical electronics component), in fact, the integrated circuit seemed to represent a departure only in terms of size. While the transistor had been considered a radically new electronic component that vertically integrated firms could not afford to ignore, moreover, the silicon-planar-integrated circuit technology was initially viewed (erroneously in the event) only as an alternative to existing materials, processes and products.

Most of these established firms were the vertically integrated receiving tubes producers that had been successful in the old germanium-alloy-mesa-discrete device technology. The technological knowledge obtained by European firms about the old technology, the capabilities accumulated in the R&D and production of semiconductor devices and the commercial success on European markets during the late 1950s and early 1960s was enough to convince these firms that the established technology was the 'right one'. Although the new technology was understood in scientific and technological terms by most semiconductor specialists and although some of the managers and technicians of the semiconductor divisions of European firms believed in the commercial possibilities of the new emerging technology, this view was contested both within the semiconductor divisions and within the management levels of the firms: germanium was considered more suitable than silicon for small signal semiconductor devices and the mesa technology was considered a very efficient and economical production process. Therefore, within the establshed semi-conductor operations of these firms, the manufacture of germanium alloy and mesa transistors and silicon controlled rectifiers continued to be the dominant activity; this manufacture eventually reached highly efficient levels. The firms concerned decided not to change their efficient production processes and commercially successful products in favour of new, highly uncertain ones. In fact, a high-level Siemens official was quoted as saying, 'There is no real demand for integrated circuits. This can be more easily met through licensing' (Scholz, 1974). Similarly, the management board at Philips entered the 1960s without any decision to begin large-scale production of silicon planar integrated circuits. Rather, they decided to confine integrated circuits to development and pilot production areas.

In general, European vertically integrated corporations were as late beginning integrated circuit production in the 1960s as they had been with transistor production in the late 1950s. As Table 5.8 shows, the delay of European producers ranged from a minimum of 1.3 years for Britain to a maximum of 3.5 years for Italy.

Table 5.8 Time lag between American and European producers in planar-epitaxial-integrated circuit production (in years)

	Great Britain	France	West Germany	Italy
Planar-Epitaxial-Integrated Circuit	1.3	3.0	2.6	3.5

Source: Tilton (1971); Freeman (1965) for Italy.

Because in the integrated circuit period technology had become increasingly cumulative, complex and appropriable, however, a delay equal to the one in the transistor period had different consequences in the integrated circuit period. Late entry during the transistor period did not greatly disadvantage the entrant; entrants that were endowed with certain technological capabilities could imitate or 'catch up' with existing producers. Late entrants during the integrated circuit period, on the other hand, were at a great disadvantage relative to existing producers; as technological change became cumulative, complex and appropriable, it became harder for firms to 'imitate' or 'catch up' with their competitors who had gained technological advantages in a product or process range.[18]

In the case of Philips, attempts to begin integrated circuit production were uncoordinated. In 1963 Philips produced an amplifier for hearing aids which was composed of three transistors and two resistors. In the following year, Philips' subsidiary in Nijmegen began to develop a digital integrated circuit for Electrologica computers. Then, in 1965, Mullard (Philips' subsidiary in Britain) produced

DTL integrated circuits for Marconi's computers; Mullard's production of DTL integrated circuits, however, was discontinued a short time later. Finally, in 1967, Philips began the mass production of digital integrated circuits. Because of its inexperience with planar technology Philips had to obtain a licence from Westinghouse in order to produce the DTL WC 200. The DTL WC 200 began to be produced in the Netherlands, Britain, West Germany and, for a very short time, Switzerland.[19]

In the case of Siemens, the beginning of integrated circuits R&D and production was stimulated by the company's computer production facilities and, later on, by the consumer market. In fact, as early as 1960, Siemens had established R&D in integrated circuits because the computer division believed that the production of integrated circuits would eventually be of some importance for computer production. However, only in the second half of the 1960s did the mass production of integrated circuits begin at Siemens; before that time, integrated circuits were considered uneconomical to produce. The first production of 'something like integrated circuits' began in the linear application area. In 1965 Siemens began the production of a three-stage amplifier for hearing aids for its medical equipment division. Then, in 1967, with a licence obtained from Westinghouse, Siemens also began producing DTL integrated circuits which were intended for use in computer production.[20]

In contrast to both Philips and Siemens, other European vertically integrated firms which were either unsuccessful in, or less committed to, the old germanium-alloy-mesa technology, were quicker to move into the production of integrated circuits during the 1960s. Being less committed to the old technology meant that they were able to appreciate more completely the advantages of the new technology as a competitive weapon against the dominant firms in the old technology. In Britain, Plessey[21] began to produce planar transistors in the early 1960s and integrated circuits in 1965. During the second half of the 1960s, however, Plessey focused its efforts on linear integrated circuits (Golding, 1971). Ferranti, on the other hand, began R&D on silicon in the early 1950s and began producing planar transistors in 1961, Micronor I DTLs in 1962 and Micronor II DTLs (a very fast DTL) in 1965 (Golding, 1971; Sciberras, 1977). In France, with the push of the French military, Cosem began producing DCTL during the 1960s. Finally, in Germany, AEG-Telefunken began R&D on integrated circuits around 1962–3 and began the mass production of DTLs, TTLs and ECLs in 1965.[22] AEG-Telefunken also obtained a licence from CBS to produce RTL integrated circuits during this period.

These European vertically integrated semiconductor firms did not grow to become major producers in the European industry as new merchant producers did in the American industry. Although these firms were ready to move into silicon-planar-integrated circuit technology, they did not switch to the new technology as quickly or as radically as American merchant producers. Nor did they commit the same amount of resources as American producers. In addition, because of the limited size of the European integrated circuit market, they were not able to benefit from either the economies of scale or the dynamic economies in integrated circuit production from which American producers benefited. Therefore, when, during the second half of the 1960s, the European integrated circuit market grew in size, American merchant producers made direct foreign investments into, and dominated, this market.

2.3 American direct foreign investments in Europe

Since most European producers were slow to react, or were not fully committed to the introduction of integrated circuits during the 1960s, American firms were able to move into the European market with exports and direct foreign investments. These American firms began the production of integrated circuits early and committed significant resources to this production. As Table 5.9 shows, during the 1960s (and particularly between 1969 and 1970), the direct foreign investments of these American producers greatly increased over earlier periods.[23]

Table 5.9 Number of known direct foreign investments in developed countries

	1950–60	1961–70	1969–70
Total developed countries	5	24	16
Great Britain	3	7	6
France	1	4	1
West Germany	1	4	3
Italy	0	3	2

Source: Finan (1975), p. 57.

At a general level, in high technology industries, firm decisions to export rather than to license technology or to invest abroad depend on the degree of internalization of firm specific advantages and on a set of other factors.[24] In high technology sectors firms' specific advantages are based on knowledge. Thus, innovative firms do not want to lose their advantage by licensing their competitors too early in the game; rather, they prefer to internalize and exploit their advantages.[25] The choice of making direct foreign investments rather than continuing to export, on the other hand, is not related to the internalization factor; firms may also be able to keep their specific advantage by exporting. Rather, this choice depends on factors such as the cost of the foreign investment, the availability of skilled labour in the source country, the cost of export marketing, the presence of tariff and non-tariff barriers to trade, the need to be near foreign customers and the degree of oligopolistic interdependence.[26]

During the second half of the 1950s few American producers made investments in Europe. As mentioned in Chapter 4, Texas Instruments, the major producer, made only one direct foreign investment in Britain in 1957 and one in France in 1960. The direct foreign investments of Texas Instruments (and, to a lesser degree, those of some of the other companies such as Hughes and International Rectifiers) were the direct result of firm-specific advantages, related to innovative activity.[27]

During the 1960s, however, because technology became increasingly complex and appropriable, the advantages of American producers over European producers grew. American producers had a general cumulative technological advantage in the new silicon-planar-integrated circuit technology and firm-specific cumulative advantages resulting from their activity in the new technology. These advantages translated initially into increasing amounts of exports and then into direct foreign investments in Europe.[28]

The specific incentives which induced American firms to make direct foreign investments in Europe varied from firm to firm, however. Texas Instruments made its first investment in France as a reaction against a similar move by IBM and because it predicted that the demand for semiconductors in Europe would grow rapidly. In addition, it wanted to be near foreign customers.[29] Signetics and National Semiconductor made their direct foreign investments in Scotland in 1969 and 1970, respectively, because skilled labour and engineers were available at relatively low costs. Motorola opened a plant in France in 1969 because of the increasing competition it faced in the market for discrete devices in Europe; France was chosen because of its large market and its high non-tariff barriers (Finan, 1975). Finally, Fairchild, a relatively new and small company, made investments in Europe because it had technological superiority in planar technology. In 1960, in fact, Fairchild acquired 33.3 per cent of the L. 2 billion ($3.2m.) capital of SGS.[30] In this deal, Fairchild was to provide SGS-Fairchild with its specific technological know-how in the planar process; at the time of the joint venture, SGS was selling germanium products and was in a highly unprofitable situation.[31] As a result of the joint venture with Fairchild, however, SGS acquired the planar technology, considerably increased its sales and, in 1963, finally showed a profit.

In addition, because the assembly stage was more labour intensive and less technology intensive than other stages of the manufacturing process, American firms also made direct foreign investments in offshore and point-of-sale assembly line processes in less developed countries during this period. Offshore assembly investments were not related to firm-specific advantages in technology. Rather, they were motivated by the need to reduce costs.[32]

European semiconductor producers followed this move of American producers a few years later, but in the meantime they were unable to match the manufacturing costs of American producers.[33] Philips opened plants in Taiwan and Hong Kong, Siemens in Singapore and Malaysia, SGS-Fairchild in Singapore, Thomson-CSF in Morocco (Truel, 1980) and Plessey in Malaysia.[34]

As a consequence, exports from offshore plants became an increasingly important source of American firms' market penetration in Europe. American firms only reimported some of the semiconductors that were assembled in less developed countries; the rest were directed to either European subsidiaries or other foreign markets. Therefore, while the 'ownership' of international trade was unaffected, its 'geography' was greatly affected by the pattern of these investments. The exports of American devices to the European market continued to increase during the late 1960s and early 1970s, while the share of semiconductors exported from American offshore assembly plants into Britain, France, Germany and Japan over the total exports of American producers into, in addition to the production of American subsidiaries in, these countries rose from 0.34 in 1968 to 0.58 in 1972 (Finan, 1975). As Table 5.10 shows, American exports from host countries to Europe increased from $46m. in 1968 to $253m. in 1972.

Table 5.10 American penetration in Britain, France, West Germany and Japan, 1968 and 1972 ($ millions)

	1968	1972
Direct exports	78	151
Value added in the 4 countries	13	33
Exports from host countries	46	253
Total	137	437

Source: Finan (1975), p. 124.

2.4 *The European defeat in the logic war*

In the late 1960s several European semiconductor producers decided to begin the production of integrated circuits on a large scale. Integrated circuits had become a commercially successful product, had a market in Europe and were increasingly used in several electronics final products. Wanting an in-house capability in custom integrated circuits and needing a secure source of supply of standard integrated circuits, the final electronics product divisions of various European corporations began putting pressure on management to initiate the production of integrated circuits in-house. As a result, the management boards of these corporations began to understand the importance of the technology and its commercial impact and finally decided to begin producing planar integrated circuit devices.

European producers, however, were not fully committed to large-scale production of standard integrated circuits of the digital type. Without extensive government support and with a demand structure requiring relatively more linear than digital integrated circuits, these European producers faced fully committed American firms, with production facilities in Europe and with technological and productive experience in digital integrated circuits. In addition, in their moves into the digital integrated circuit market, these European producers chose the wrong technological partners when they decided to purchase licences. In fact, they obtained licences from American vertically integrated receiving tube producers, the losers in the integrated circuit race, and not from American merchant producers, the winners.

The technological and productive strength of American producers was complemented by two major events—the bipolar logic war and the semiconductor recession of 1970–71. These events militated against the successful entry of European producers into the standard digital integrated circuit market.

The TTL logic war[35] that occurred during the second half of the 1960s produced a clear winner—Texas Instruments—in the digital integrated circuit market. TTL bipolar integrated circuits were the standard components used in the computer industry in the mid-1960s. Fairchild (a new merchant producer) had become the American industry's leader in RTL and DTL digital integrated circuits. Fairchild's DTL F930 had several second sources and was expected to remain the technological standard for several years. However, during the second half of the 1960s, Sylvania (a vertically integrated receiving tube producer) introduced a new type of bipolar integrated circuit, the TTL,[36] and Texas Instruments (a new merchant producer) soon followed Sylvania's lead in TTL production. Instead of focusing on design, however, Texas Instruments decided to attempt to increase the integration of the functions of the device. After aggressive price cutting[37] and after National Semiconductors (a new merchant producer) became a second source for Texas Instruments, the 54/74 TTL of Texas Instruments became the industry leader in the late 1960s and early 1970s; its lead, in fact, was further reinforced after it was coupled with Texas Instruments' high-speed Schottky TTL in 1970 and with the lower power Schottky TTL in 1973. (Table 5.11 shows the market shares of Texas Instruments in 1970.) The other major competitors in the TTL market, Sylvania and Fairchild, were less successful than Texas Instruments. Sylvania, in fact, abandoned the semiconductor industry in the 1970s and Fairchild did not introduce its TTL 9000 until much later (Wilson, Ashton and Egan, 1980).

Table 5.11 Market shares of Texas Instruments, 1970

13% of world market for semiconductors[a]
17% of world market for integrated circuits[a]
25% of European integrated circuit market[b]
50% of European TTL market[c]

Sources: [a]'Dataquest' in Dosi (1981); [b]*Business Week*, 3 January 1970; [c]*Business Week*, 22 August 1970.

The bipolar logic war partially overlapped with the second event, the 1970–71 semiconductor recession. This recession was linked to the overall behaviour of the economy, and was characterized by a decrease in demand for semiconductor devices. The decrease in demand spurred the competition among existing producers (Table 5.12).

Table 5.12 Firm shares of the semiconductor market in Europe, 1969
(percentage of total market)

	A %	B %
Philips	22.5	30
Texas Instruments	15	18
SGS	9	8
Siemens	8.5	7
ITT	7.5	7
Sescosem	7.5	na
Motorola	5	9
Telefunken	2.5	3
Ates	2	na
RCA	2	na
Silec	2	na
Transitron	1.5	na
Ferranti	1.5	na
Others	13.5	18
Total	100	100

Note: In 1969 15 per cent of the European semiconductor market was in-house.
Sources: (A) SGS; (B) *Electronic News*, 27 June 1970 in Webbink (1977), p. 32.

The bipolar logic war and the semiconductor recession provoked a product winner (the TTL 54/74 of Texas Instruments) and a major reduction in prices. In 1970, the price of a TTL circuit decreased from $0.6 (January 1970) to $0.15–0.2 (August 1970) in Britain, and from $0.45 (February 1970) to $0.3 (August 1970) in France. In West Germany the price of 4 Gates and 20 Gates TTL fell respectively from approximately 4DM and 19DM to 2DM and 12DM.[38]

As a consequence of the commitment of American firms, the logic war and the semiconductor recession, European semiconductor producers were unsuccessful in their attempt to begin large-scale production of digital integrated circuits. In a growing, though still limited, market with few producers (such as the integrated circuit market was during the first half of the 1960s), a firm could enter the market late and still be successful. However, with cumulative technological change and a market composed of several producers aggressively competing in the midst of a crisis, as was the case in 1970–71, late entrants could be successful only if they committed large amounts of investment. The following analysis of the history of the major European producers will examine the various cases in more detail.

Philips, for example, invested resources in integrated circuits, but dispersed its production over a range of different products without effectively integrating them at the product level or with the production of its electronics final goods. As previously examined, Philips decided to begin the production of WC 200 DTL integrated circuits on a large scale in 1967.[39] Part of Philips' production of WC 200 DTL went into the computers that were produced by Electrologica, a Philips subsidiary.[40] However, Electrologica's computers were not commercially successful and therefore Philips decided to switch from the production of WC 200 DTL to Fairchild's 930 DTL. In addition, during the late 1960s, Mullard (Philips' British subsidiary) began producing 7400 TTL circuits for GEC (GB),[41] while Radiotechnique (Philips' French subsidiary) chose to produce Sylvania's SUHL-TTL devices, which were sold in large quantities to Bull (France) and to Philips' computer division. Philips' Swiss subsidiary decided to develop and produce optical devices for camera producers such as Kodak and Polaroid during the late 1960s and early 1970s. Finally, in the late 1960s, Philips developed ECL and I2L bipolar integrated circuits and MOS integrated circuits for desk calculators; these lines, however, were never very successful.[42]

Table 5.13 Semiconductor patents in Great Britain by firm (shares of total patents)

	1961–4	1965–7	1961–7
Siemens	10.3	7.5	8.9
Siemens-Halske	5.5	2.6	4.0
Philips	9.0	7.0	8.0
Mullard (Philips)	2.2	1.7	1.9
ITT (STL)	5.0	10.9	8.0
IBM	3.3	9.4	6.5
West E.	6.4	4.6	5.4
Philco	9.3	1.8	5.4
AEI	6.3	2.6	4.4
Texas Instruments	4.3	4.1	4.2
WE	4.0	3.9	3.9
RCA	2.3	3.9	3.1
GEC	3.8	1.1	2.4
AEG	0.5	1.1	0.8
Telefunken	1.5	1.7	1.6
West Brakes	1.7	2.3	2.0
Motorola	0.2	2.5	1.4

Source: Golding (1971).

114   *The integrated circuit period*

Table 5.14 Semiconductor patents in France by firm (share of total patents)

	1961–4	1965–8	1961–8
Siemens	26.5	28.1	27.4
Philips	20.3	18.1	19.1
IBM	8.2	14.2	11.5
Texas Instruments	4.9	7.7	6.4
ITT	4.3	6.1	5.3
Thomson-Houston	7.0	3.5	5.0
AEI	6.4	2.7	4.3
AEG-Telefunken	2.3	4.9	3.7
Motorola	1.9	4.0	3.0
Westinghouse Brake	3.5	1.3	2.3
CSF	2.6	1.1	1.7
CGE	1.1	1.8	1.5
Emihus	1.2	1.0	1.1
SGS	1.6	0.6	1.1
Intermetall	2.0	0.4	1.1

Source: Tilton (1971).

The other major European semiconductor manufacturer, Siemens, had a more co-ordinated approach to integrated circuit production; however, it was equally unsuccessful in the competitive production of bipolar devices. In the period 1967–70 Siemens produced limited quantities of TTL, mainly for use in its 4004 computers; these TTL were special types (distortion independent), created because the computer division of Siemens had asked for such devices.[43] As a consequence, when Texas Instruments slashed its prices for TTL devices in 1970–71, Siemens decided not to enter international TTL competition. Rather, it chose to purchase TTL devices on the open market.[44]

Philips and Siemens, however, were not the only major European semiconductor manufacturers to be unsuccessful in the production of TTL integrated circuits; AEG-Telefunken was unsuccessful as well. In the mid-1950s, AEG-Telefunken developed DTL, TTL, and ECL circuits.[45] The TTL circuits were licensed from Sylvania while the ECL were developed for fast computers.[46] AEG-Telefunken's digital bipolar devices were produced in small quantities, however, and soon discontinued. AEG-Telefunken exited from digital bipolar integrated circuit production partly because it lacked a source of internal demand; its computer division (and its subsidiary Olympia) did not purchase enough TTL devices from the semiconductor division to permit the exploitation of scale economies. As a result, AEG-Telefunken decided to turn its efforts toward linear and custom integrated circuits[47] at the end of the 1960s.

While Philips, Siemens and AEG-Telefunken obtained a limited share of the digital integrated circuit market, they maintained high market shares (albeit declining) in the overall semiconductor market. As Table 5.15 shows, in the German semiconductor market in 1968, Valvo (Philips) held 25 per cent of the market, Siemens 22 per cent and AEG-Telefunken 9 per cent.[48] These firms remained, in fact, highly competitive in discrete devices and in linear integrated cicuits.[49]

Table 5.15 Firm shares of the West German semiconductor market, 1968 (percentages of total market)

	%
Valvo (Philips)	25
Siemens	22
Texas Instruments	16
Intermetall (ITT)	10
AEG-Telefunken	9
SGS	6
Others	2
Imports	10
(of which Motorola had	4)
Total	100

Note: The shares of American firms in the German semiconductor market grew from 36 per cent in 1968 to 51 per cent in 1971 according to Finan (1975).
Source: Tilton (1971), p. 115.

Like Philips, Siemens and AEG-Telefunken, most British producers did not have commercial success in the manufacture of bipolar integrated circuits either.[50] STC, Plessey, and Lucas absorbed between 50 and 80 per cent of their semiconductor production in-house; their focus was on discrete devices and linear integrated circuits. AEI tried to re-enter the semiconductor industry through the production of planar transistors and integrated circuits, but failed. GEC, on the other hand, did not commit large resources to digital integrated circuit production because it was not deeply involved in semiconductor operations.[51] Finally, during the 1960s, English Electric produced only limited quantities of digital integrated circuits through its Marconi and Elliott subsidiaries. In the early 1960s Marconi produced multichip digital integrated circuits for its process control computers and in 1966 it manufactured the Micronor II DTL with a licence from Ferranti. However, sales on the open market remained limited and, as a consequence, the focus of Marconi's production shifted to custom devices for internal use. Instead of relying on Ferranti's licences, Elliott Automation produced standard devices; it produced the 930 DTL and the 990 TTL with licences from Fairchild. With the defeat of Fairchild in the TTL logic war, however, Elliott Automation's production of digital integrated circuits was negatively affected.

Therefore, as in the German case, domestic producers had a limited share in the British integrated circuit market while they maintained a relatively large share (although declining) in the overall British semiconductor market. As Table 5.16 shows, while in 1967 Texas Instruments, SGS-Fairchild and Motorola held 55 per cent of the British integrated circuit market, they held only 43 per cent of the British semiconductor market. Compared to 1962, however, these firms increased their market shares more than 30 points (from 13 to 43 per cent).

Table 5.16 Firm shares of the British semiconductor market, 1962 and 1967 (share of the total market)

	British semiconductor market		British integrated circuit market
	1962 %	1967 %	1967 %
ASM (Philips-GEC)	49	23	1
Texas Instruments	13	22	25
Ferranti	10	5	6
AEI	7	4	−
Westinghouse Brake	5	5	−
STC (TT)	2	7	7
SGS-Fairchild	−	16	21
Motorola	−	5	9
Marconi-Elliott	−	2	11
Others	14	11	20
Total	100	100	100

Notes: In-house sales are included.
The market share in semiconductors of American firms grew from 53 per cent in 1968 to 59 per cent in 1972 according to Finan (1975).
Source: Golding (1971), pp. 179–80.

In the British semiconductor industry, one major exception to this situation existed: Ferranti has been the most flexible, adaptive and technologically progressive of the British semiconductor firms. In 1962 Ferranti developed the Microlin and Micronor I devices for small computers and in 1965 it developed the high-speed Micronor II DTL (based on RCA technology) for Ferranti's process control computers. However, neither the computer nor the military market absorbed large quantities of the Micronor II DTL. Then, in the second half of the 1960s, Ferranti developed its own TTL devices. During the period 1966–70 Ferranti sold large quantities of TTL devices to the British computer producer ICL, because Texas Instruments was mainly serving the booming American market. However, with the bipolar logic war and the semiconductor recession, ICL once again began purchasing large quantities of Texas Instruments'

57/74 TTL instead of Ferranti's TTL. Because of the lack of demand for its TTL devices, Ferranti had to abandon the standard TTL market. Ferranti also had weaknesses, however. It lacked sufficient resources for financing a full-scale effort in R&D in digital integrated circuits; it kept the semiconductor division dependent on the policies of the other divisions within the firm; and it focused its R&D on engineering development and cost reduction, with less emphasis on marketing considerations.[52]

Similarly to the German and the British cases, the domestic production of integrated circuits also remained limited in the French semiconductor industry.[53] Cosem (a subsidiary of CSF)[54] was the major producer of bipolar digital integrated circuits, but its absolute production remained limited and covered only a minor part of Cosem's overall semiconductor production. The same is true for Sescosem, which was particularly successful in zener diode production.[55]

The French case is therefore similar to the German and British cases; French firms had limited shares in the French integrated circuit market although they maintained a large share (although declining) of the overall French semiconductor market. As Table 5.17 shows, Sescosem held 20 per cent of the French semiconductor market in 1968.

Table 5.17 Firm shares of the French semiconductor market, 1968 (shares of total market)

	%
Radiotechnique Compelec (Philips)	22
Sescosem	20
Texas Instruments	20
Silec	7
SGS	7
Motorola	5
Others	4
Importers	15
(of which Intermetall had	3)
Total	100

Note: The market share in semiconductors of American firms grew from 33 per cent in 1968 to 55 per cent in 1972 according to Finan (1975).
Source: Tilton (1971), p. 115.

The Italian case is very different from that of the other European countries, because SGS-Fairchild was highly successful in digital integrated circuits. SGS-Fairchild was a multinational corporation with 20 per cent of the Italian market in 1973, as Table 5.18 shows.[57]

In the period 1961-8, SGS-Fairchild became a leader in planar technology and increased its market share in Europe. SGS-Fairchild began producing Fairchild

Table 5.18 Firm shares of the Italian semiconductor market, 1973 (shares of total market)

	%		%
SGS	20	Philips	7
Texas Instruments	15	ITT	7
Fairchild	13	Siemens	6
Motorola	10.5	Others	21.5

Source: Pertile (1975).

RTLs in 1962 and Fairchild DTLs in 1964. These products were addressed to the computer, professional and, to a lesser extent, military markets. In fact, in the early 1960s Olivetti was the major customer of SGS-Fairchild.[58] However, Olivetti's captive market was greatly reduced in 1964 when Olivetti sold its computer division to General Electric. With the introduction of silicon planar technology in the 1960s, SGS-Fairchild increased its market share in Europe. It opened foreign subsidiaries in Britain (1963), France (1966), West Germany (1966) and Sweden (1966). Rather than specialize in different products, however, these plants chose to produce basically the same product lines; SGS-Fairchild decided to have facilities that were producing a wide range of products near their potential customers.

The technological dependence of SGS-Fairchild on Fairchild gradually declined during the 1960s, however. In fact, the management of SGS-Fairchild realized that the European semiconductor market was consumer-orientated and that there was a growing demand for custom and speciality devices. These were neither the markets nor the types of products that their American partner was addressing; Fairchild's R&D was done in the United States and was aimed at the development of standard devices for the computer and industrial markets. The divergence of targets (custom-speciality devices and consumer market for SGS, standard devices and computer market for Fairchild) led the Italian SGS-Fairchild management to create a separate R&D group in Europe during the mid-1960s. This new group was assigned to do basic and applied research on digital and analog devices. This independent R&D effort gradually separated SGS-Fairchild from Fairchild's strategies, both in terms of final markets and in terms of the types of products that were being developed.[59]

In 1968 Fairchild divested itself from SGS for two reasons. First, Fairchild was experiencing both technological and financial difficulties in the United States (Golding, 1971; Wilson, Ashton and Egan, 1980). Second, it was unable to compel SGS-Fairchild to produce specific types of devices (digital integrated circuits) for specific final markets (the computer market). After the withdrawal of Fairchild, SGS and Fairchild continued to exchange technological know-how, however, and SGS continued to produce the Fairchild 900 TTL.[60]

The withdrawal of Fairchild meant that SGS was formally a vertically integrated corporation. Olivetti, in fact, had complete control over SGS.[61] However, vertical integration was ineffective because Olivetti did not absorb large quantities of devices produced by SGS. By the late 1960s Olivetti had, in effect, almost completely abandoned the electronics sector.

After the withdrawal of Fairchild, the position of SGS in the digital bipolar market gradually declined. This reduction occurred for three reasons. First, SGS was affected by the defeat of Fairchild in the TTL war and by the victory of Texas Instrument's 74/54 TTL. Second, SGS did not commit large amounts of resources to R&D in standard digital bipolar integrated circuits. Rather, SGS focused its product choices on linear and power integrated circuits and on custom-speciality devices. Third, SGS did not receive as general and complete a transfer of technology from Fairchild after the split as it had in the first half of the 1960s; SGS was considered a potential competitor of Fairchild in the European market, especially after Fairchild decided to establish its own plants in Europe.[62]

In conclusion, by the beginning of the 1970s, all of the major European semiconductor producers had been pushed out of the standard bipolar digital integrated circuit market. Most European corporations decided instead to concentrate their productive efforts on discrete devices, linear integrated circuits or custom and speciality devices.

3 Supply factors: the decline of vertically integrated receiving tubes firms

Vertically integrated receiving tubes producers continued to dominate the European semiconductor industry during the 1960s. These producers included Philips, Siemens, AEG-Telefunken, Thomson-CSF, GEC.

These producers began doing R&D and producing small quantities of integrated circuits for a set of incentives similar to those which receiving tubes producers had followed to begin the production of transistors. All these producers were in fact transistors producers: they could use the technological capabilities which they had accumulated in the R&D and production of discrete devices and could, therefore, reap economies of scope in the R&D and production of discrete devices and integrated circuits. Moreover, because integrated circuits could be substituted for transistors in many applications, they would be able to sell these integrated circuits to their transistor customers.

These electronics final goods producers, however, were also given an additional incentive to integrate vertically by the new interrelatedness of the designs of integrated circuits and electronics final products (Figure 5.2). During the 1950s the transistor had been placed in an electrical circuit which was part of a functional apparatus that was, in turn, part of a system. As a result, the physicist was separated from the engineer, who was in charge of the development

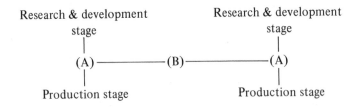

INTEGRATED CIRCUIT   ELECTRONIC FINAL PRODUCT

Research & development   Research & development
        stage   stage
          | |
        (A) ——————— (B) ——————— (A)
          | |
   Production stage   Production stage

Figure 5.2 Integration of integrated circuits and electronic final products

of the device, and from the user of the component. During the 1960s, however, this situation changed. The integrated circuit was at the same time a component and a circuit, and was part of a functional apparatus. Therefore, physicists and engineers had to work in close contact with one another and had to be well informed about the eventual uses of the devices they were developing. In addition, since technological advances were increasingly based on productive and engineering capabilities, R&D staff needed to be better informed about developments at the manufacturing stage. As a result, the design of circuits moved upstream to the component level. In particular, the designs of custom-integrated circuits were tailored to specific customers for specific uses; they had to be produced through co-operation and close contact between component producers and users.

This incentive, however, had to be coupled with a set of other factors in order to determine the vertical integration of the R&D and production of electronics final goods with R&D and production of semiconductors. While the R&D and production in semiconductors or in final goods (*A* of Figure 5.2) had already been integrated 'by tradition' even before the transistor period, the integration of the R&D and production of integrated circuits with that of final goods occurred for a different set of reasons. At a general level, in fact, co-operation at the design stage between an upstream and a downstream product implies co-operation in R&D, but does not necessarily imply vertical integration of the R&D and the production stages.[63]

In the semiconductor industry these factors can be subdivided either into factors affecting standard or into factors affecting custom integrated circuits. Such factors did not exert their effects when the cost of producing integrated circuits in-house was much higher than the price of buying them on the open market.

In the case of standard integrated circuits, the size of the internal market, the eventual profitability of external sales and the uncertainty of the supply of components provided general incentives for upstream vertical integration. The size of the internal market and the uncertainty of supply, for example, played an important part in IBM's decision to integrate vertically in the early 1960s.

In the case of custom integrated circuits, the long-run effects of the in-house production of components on innovation at the final goods level, and the transaction costs faced by electronics final goods producers provided general incentives for vertical integration. These transaction costs were particularly high, the more innovative and complex the custom integrated circuits purchased and the smaller the quantities of devices required. In this case, in fact, merchant producers would face high investments and set-up costs and therefore were often reluctant to begin such production. In the case of custom components which are needed on a continuous basis for a long period of time, an 'implementation problem' may arise. This problem involves the adjustments of plans and a continuous co-ordination among firms.[64]

Moreover, electronics final goods producers would try to avoid the short-run effects of possible disclosures of proprietary information about their products by the merchant producer. Such disclosures may occur in two situations. First, if a merchant producer decides to begin the production of large quantities of a standard integrated circuit (rather than continue to produce small amounts of custom-integrated circuits), it could use the orignal designs of the custom integrated circuit as a basis of the standard. Second, merchant producers could sell or exchange licences for custom integrated circuits.[65]

In Europe, these specific incentives for vertical integration in integrated circuits R&D and production worked only for custom integrated circuits and were carried on by the existing vertically integrated receiving tubes producers. These producers, in fact, were neither large enough to absorb a large share of the amount of standard integrated circuits that would be produced in-house, nor innovative enough to be able to be commercially successful in the external standard integrated circuit market.[66] On the other hand, no new major European electrical final goods producer integrated vertically upstream into the R&D and production of integrated circuits.[67] These producers had a limited in-house demand for semiconductors, were unwilling to take risks in producing new products, and preferred to buy state-of-the-art integrated circuits from the external market.[68]

The European vertically integrated receiving tube producers which began R&D and production in integrated circuits in the mid-1960s did not commit themselves to integrated circuits production in the early 1960s, because of technological beliefs, or to digital integrated circuit production in the late 1960s, because of corporate focus on electronics final goods. These producers, in fact, regarded the integrated circuit as an alternative to the transistor, with limited applications. In addition, corporate focus on electronics final goods meant that the semiconductor divisions were not supported with sufficient resources or a strong commitment in their efforts to produce digital integrated circuits successfully. Support was given, on the other hand, to the R&D and production of discrete devices and linear integrated circuits which were widely used in the electronics final goods of these corporations.

A strong commitment to digital integrated circuit production would have been possible either by firms which produced only these types of devices (merchant manufacturers) or by firms which used them extensively (computer producers). Both types of firms were absent in Europe during the 1960s. New European merchant producers did not appear on the scene. European computer producers continued to be limited in size and unsuccessful compared to the American computer producers. This last aspect will be examined in the next section, on the demand structure of the European semiconductor industry.

4 Demand factors: the lack of a competitive European computer industry

During the integrated circuit period, as during the transistor period, the electronics consumer goods market continued to be the largest semiconductor end-user market in Europe. As Table 5.19 shows, the electronics consumer goods market had a share of 31 per cent of all electronics final markets in Europe, vs. less than 18 per cent in the United States.

Table 5.19 Electronics final markets in the United States and in Europe, 1966–1970 (percentage of total electronics final markets)

	United States	
	1966	1970
Consumer	18	14
Communication	7	7
Computers	13	24
Industrial-Med.-Test	11	8
Federal[a]	51	47

	Four Countries (Britain, France, Germany, Italy)	
	1966	1970
Consumer	31	31
Communication	25	24
Computers	19	22
Industrial	25	23

Note: [a]Federal includes electronic components
Source: *Electronics*: various years

The relative importance of the consumer market in Europe implied a demand for a well-defined type of semiconductor. While the computer/calculator market in fact required mainly digital integrated circuits, the consumer market required relatively more discrete semiconductors and linear integrated circuits.

The relative importance of the consumer market in Europe also meant that electron tubes continued to be greatly in demand. As Table 5.20 shows, while the computer/calculator and military production had a low electron tube content, consumer electronics had a high electron tube content.

Table 5.20 Semiconductor content of electronic final products

	Year	Semiconductor (as % of all materials)	Electron tubes (as % of all materials)
Military and industrial electronic equipment	1967	7	2
Computer/calculator	1967	8	0
Telephone and telegraph equipment	1967	5	1
Consumer electronics equipment	1967	2	29
Electrical test and measuring equipment	1967	6	2
Engineering and scientific instruments	1967	1	1

Source: US Department of Commerce (1979), p. 34.

The relative importance of the consumer market in Europe was coupled with the competitive strength of European corporations in both the consumer and the industrial markets. Philips, Siemens, Thomson-CSF, AEG-Telefunken, GEC and Ferranti were highly competitive in one or both of these markets.

During the integrated circuit period also therefore the consumer and industrial markets influenced the evolution of the European semiconductor industry by exerting a strong demand for discrete devices and linear integrated circuits. Linear integrated circuits were difficult to standardize and firms were therefore able to create and occupy market 'niches' by producing speciality devices.[69] In turn, these market niches were relatively immune from worldwide competition and permitted the firms which occupied them to survive in the market.

The influence of these markets is reflected in the early commencement of production of linear integrated circuits in most European countries. In West Germany (which had a large consumer and industrial market), for example, the large-scale production of linear integrated circuits began before the large-scale production of digital integrated circuits, in most cases. Siemens, in fact, began the mass production of linear integrated circuits in 1965 and digital integrated circuits in 1967, while Intermetall (ITT) began the production of linear integrated circuits in 1967. The only exception was AEG-Telefunken, which began to produce digital integrated circuits in 1965 and linear integrated circuits in 1967 (Scholz, 1974, p. 139).

European semiconductor producers reacted differently to the pull of the consumer and industrial markets. For several of these corporations this pull was internal, while for others it was external, through the open market.

Philips incrementally innovated in linear integrated circuits for television, because of the pull of in-house customers. Philips produced diodes, transistors and power devices for its consumer and industrial products and linear integrated circuits for its consumer products. In general, approximately half of Philips's in-house production of semiconductor devices was used in Philips's products, and more than half of the semiconductor devices required by Philips came from in-house sources. In the late 1960s and early 1970s the consumer division of Philips 'pulled' the development of an analog integrated circuit for televisions: this development arose because in-house television producers had realized that they needed integrated devices for televison receivers.[70]

AEG-Telefunken and Thomson-Brandt used semiconductors produced in-house in their electronics consumer goods products. In the second half of the 1960s, AEG-Telefunken focused its semiconductor production on discrete devices and linear integrated circuits because it had a large internal consumer market.[71] The Thomson group,[72] through Sescosem, produced discrete devices and linear integrated circuits, a share of which were used in-house.

Siemens, Plessey and Lucas, on the other hand, produced discrete devices and linear integrated circuits to be used in industrial equipment.[73] However, the position of Siemens differed from that of Lucas and Plessey:[74] Siemens was a major producer of semiconductor devices for the external consumer market. In the second half of the 1960s, in fact, Siemens' sales of devices to the external market exceeded the value of semiconductors used in-house and in 1969, one-third of Siemens' production of components[75] was sold to the consumer market, most of which was external to the firm (*Siemens-Berichte*, 1969). Moreover, because of Siemens' involvement in computer production and because of the setbacks it experienced during the bipolar logic war (late 1960s), Siemens' purchases of semiconductors from external sources were approximately equal to its use of semiconductors produced in-house.[76]

The European market for semiconductors for telecommunication equipment remained limited during the 1960s (because neither the size of the telecommunication equipment markets nor the semiconductor content of telecommunication equipment increased) and therefore exerted a limited influence on the rate and direction of technological change in the European semiconductor industry. The telecommunication equipment market maintained a pattern of quasi-vertical integration between the common carriers and the telecommunication equipment producers. Yet the common carriers in most cases continued to have conservative attitudes towards new types of equipment; these attitudes translated into a lack of innovative stimulus at the semiconductor level through the public procurement of telecommunication equipment. In countries like Germany, in fact, this attitude was reinforced by the limited involvement of the common carrier in the R&D process of telecommunication producers (Diodati, 1980).

In addition, the size and characteristics of European public procurement did not change much during the 1960s: public procurement was small in size[77] and was directed towards few established firms. As Table 5.21 shows, in 1966 public procurement ranged from a maximum of 15-20 per cent in Britain, to negligible amounts in West Germany and Italy.

During the 1960s the European computer market exerted a well-defined influence on the rate and direction of technological change in the European semiconductor industry, by not requiring large amounts of digital integrated circuits. This limited demand for digital devices produced by European firms was a consequence of the limited size of the European computer market, of the continuous lack of success of European computer producers, of the domination of American computer producers, and of the failure of European government policies in support of the domestic computer industries. The following pages analyse the effects of these intersectoral linkages and interdependencies on the semiconductor industry in more detail.

Table 5.21 The size of government markets, by country, 1966 (government market as a percentage of semiconductor output)

Country	1966 %
Great Britain	15–20
France	12 (1964)
Italy	5 (for all electronics components)
West Germany	negligible
Japan	negligible

Note: In 1972, the military markets accounted for 24 per cent of the total electronics markets in the United States and 14 per cent in Europe (Finan, 1975 in Braun and Mac-Donald, 1982).
Sources: OECD (1968), p. 26 and Golding (1971), p. 141 for Great Britain.

During the 1960s the European computer market was not as large as the American computer market. As Table 5.22 shows, the European market was never greater than one-quarter of the American market.[78]

Although European vertically integrated computer firms were involved in computer production they did not try (and if they tried they were not successful)

Table 5.22 Relative size of European (EEC and Britain) computer markets
1963 and 1967
(American market = 100)

	1963	1967
Average monthly rental	17	23
Number of machines installed	20	24

Source: OECD (1969).

Table 5.23 Market shares of major computer producers, by country, 1967

	1967 Number of computers installed (%)	1967 Value of computers installed (%)
United States		
IBM	50	na
Sperry Rand	12	na
France		
IBM	43	63
Bull/GE	31	20
Great Britain		
ICL	42	45
IBM	29	39
NCR-Elliott	13	4
West Germany		
IBM	55	68
Siemens	4	9
Telefunken	1	na
Philips/Elettrol.	1	na
Zuse	8	na
Italy		
IBM	na	66
GE-Olivetti	na	19

Note: In 1966 the share of foreign subsidiaries in the production of computers of various countries was 15–20 per cent in Britain, 85–90 per cent in West Germany, 90 per cent in France, 100 per cent in Italy (OECD, 1969, p. 79).
Source: OECD (1969).

to enter the market for standard general purpose computers. As Table 5.23 shows, American producers (in particular IBM) dominated most European markets.

Philips, for example, tried to enter the computer industry by acquiring Electrologica (in 1960) and part (40 per cent) of Simag (West Germany), and by introducing its first computers in 1965. However, it did not make a full-scale entry into computer production.

Siemens' effort in computer production was more substantial than that of Philips. Its major computer products consisted of the 3003 computer, which was introduced in 1965, and the 4004 computer, which was based on a licence from RCA.[79]

During the 1960s the other German computer manufacturer, AEG-Telefunken, also had a small share of the German computer market:[80] AEG-Telefunken produced a few large specialized computers with a licence from RCA and, like Siemens, developed ECL integrated circuits for fast computers.

In Britain domestic firms had a larger share of the market than did American firms; this was very different from the German case, described above.[81] However, during the 1960s the market share of American firms in Britain increased.

The French computer industry had been penetrated by American firms much earlier than the British computer industry. Bull, for example, the largest domestic producer, obtained a licence from RCA for a medium-size computer, the Gamma 30, in the early 1960s. In 1964, however, Bull was acquired by General Electric.[82] As a consequence, 90 per cent of the French computer market came to be supplied by American controlled firms (OECD, 1969).

In Italy, the exit of Olivetti from computer production in 1964 and the absence of any public policy toward the electronics industry during the 1960s meant that there was no large domestic computer producer to 'pull' the domestic semiconductor industry. Olivetti, in fact, remained mainly a mechanical and electromechanical office equipment producer until the early 1970s,[83] although it did introduce one major electronics products, the Programma 101, in 1965: this was the first electronic desk calculator and was very successful.[84]

European governments tried to increase the competitiveness of domestic computer producers, but their attempts were unsuccessful. Between 1967 and 1970 the first German Data Processing Program was begun through the Ministries of Economics and Education and Science. This program granted DM 386.6m. ($96.6m.) to support the R&D of hardware manufacturers, computer applications, basic research and data processing education. Only DM2.2m. ($550,000), however, was given to support R&D in electronics components (1969-70).[85] In 1964 the British government financed the 'Advanced Computer Technology Project', with a grant of £5m. and supported 50 per cent of all basic research.[86] In addition, in 1968 the government supported the constitution of ICL. The French Plan Calcul supported the constitution of a domestic computer producer, CII, through a merger of CAE (a subsidiary of CSF and CGE) and SEA (a subsidiary of Schneider).

Table 5.24 R&D in the semiconductor industry by country, 1965 and 1968 ($ millions)

	1965	1968
United States	90	na
Germany	na	na
France	9	13.5
Great Britain	na	10
Italy	1	na

Note: na = not available
Sources: OECD (1968), p. 176 for 1965 and Tilton (1971), p. 129 for 1968.

Vertically integrated European computer firms certainly produced digital integrated circuits in-house, but because they did not aim at, or were unsuccessful in, the standard general purpose computer market, they produced digital integrated circuits in limited quantities. Both Philips's R&D in digital integrated circuits and its choice of producing the Westinghouse WC 200 DTL stemmed from its decision to produce devices which could be used in Philips' computer production.[87] However, because of its limited computer production, a large internal market for digital devices could not develop. The same is true for Siemens and AEG-Telefunken, which developed ECL integrated circuits for fast computers. Moreover, in the case of English Electric and Ferranti a large in-house demand for digital integrated circuits failed to materialize. English Electric produced part of the components used in its computers in-house, under a licence from Fairchild. Ferranti introduced the Microlin, Micronor I and Micronor II DTL and produced part of the components for its Argus process control computer for military applications in-house.

The other non-vertically integrated European computer producers such as ICL and CII purchased state-of-the-art digital devices from American producers rather than purchasing them from European producers, or trying to produce them in-house. The purchase of digital integrated circuits from European firms or in-house production would merely have added another level of weakness to their computer production and would have given American producers another advantage. A manager of a European firm emphasized that European corporations could not try to be competitive with both computer leaders (IBM) and semiconductor leaders (Texas Instruments). For example, according to the French Plan Calcul Sescosem (Thomson-CSF) was to produce TTL integrated circuits for the domestic computer producer CII. However, CII did not buy this type of integrated circuit from Sescosem: CII was France's 'national champion' and in order to compete with IBM in computers it needed to purchase the best 'state-of-the-art' integrated circuits from American producers.[88]

For the same reasons, vertically integrated European producers in several instances also chose to complement their internal production of semiconductors by purchasing state-of-the-art components externally. For example, Olivetti considered SGS as only one of its suppliers of semiconductors; it gave no preference to the 'domestic' producer.[89]

As a consequence, the demand for digital integrated circuits of European producers remained limited. When TTL circuits were introduced and the logic war began, most European computer producers (both vertically integrated and non-vertically integrated) adopted standard American TTL devices, in order to protect their (albeit weak) position in the computer market.

5 Public policy factor: the continuous lack of support

The character of government R&D and production refinement support in Europe did not change appreciably from the 1950s to the 1960s; it continued to be concentrated more on research than on development and to channel funds into a few, well-established domestic producers. Although the amount of support greatly increased it did not reach large amounts in absolute terms. As Table 5.25 shows, in 1968 government funding of R&D expenditures as a percentage of total R&D expenditures ranged from a maximum of 35 per cent for Britain to negligible amounts for West Germany and Italy.

Table 5.25 Government funding of R&D expenditures of firms in the semi-conductor industry by country, 1968

	In value 1968 ($ millions)	In percentage of the total R&D in the industry %
Great Britain	3.5	35
France	3.6	27
Germany	negligible	—
Italy	0	—
Japan	0	—

Sources: OECD (1968), p. 178 and Tilton (1971), p. 129.

In West Germany support for the semiconductor industry went to Siemens, Valvo (Philips) and AEG-Telefunken and was part of several different programmes. The Federal Ministry for Research supported the R&D of electronic components under a programme entitled 'New Technologies'.[90] This programme provided approximately DM2.2m. ($564,102) in grants in 1969–70 for activities in opto-electronics components, basic developments and new components, but nothing for integrated circuits.[91] It covered up to 100 per cent of the cost of overall research in non-industrial research institutions and up to 50 per cent of the overall cost of R&D in firms (BMFT, 1974); a greater share of R&D costs was covered in cases of extremely high risk and in cases which concerned strong public interest. In addition, the German Data Processing Program (1967–70), for example, gave DM8.2m. ($2.1m.) to support R&D in integrated circuits and DM5.8m. ($1.5m.) to support R&D in memory devices in 1969–70. The space programme of the Federal Ministry for Research gave DM2.6m. ($650,000) to R&D in semiconductors and DM1.3m. ($325,000) to R&D in integrated circuits, while the Federal Ministry of Defence funded R&D with DM11.2m. ($2.8m.). Finally, the Deutsche Forschungs Gemeinschaft[92] was provided with DM3.3m. ($825,000) for R&D in semiconductors and with DM10.7m. ($2.7m.) for basic research (Federal Republic of Germany, BMFT, 1974).

The British government was more 'interventionist' than the German government because it not only supported the semiconductor industry, but also promoted mergers among the domestic firms in the industry. As in the German case, the object of government policy towards electronics during the1960s was the computer industry, not the semiconductor industry. Of the £5m. allotted to the Advanced Computer Techniques Project since 1964 (Dosi, 1981), only a small fraction (£280,000 a year) has gone to finance R&D in semiconductors (Golding, 1971, p. 360). In addition, the government supported the merger of GEC and AEI in 1967 and of GEC-AEI and EE in 1968. Through these mergers the semi conductor activities of these firms were consolidated.

The CVD[93] was the major source of government R&D support for the semiconductor industry in Britain. In 1968 the CVD financed £1.35m. of the industrial R&D in this field; this amount, however, represented a decline in real terms from the 1950s. During the 1960s Plessey became the main CVD contractor for integrated circuits;[94] Plessey received support for R&D in linear integrated circuits, which it supplied to the British military. In 1962 Ferranti also received funds from the CVD to begin an integrated circuit programme from which the Micronor I was developed.[95]

Other agencies in Britain included the Ministry of Technology, the National Research Development Corporation, the RRE, the SERL and the Post Office Research Station. In 1968 the Ministry of Technology awarded Ferranti and Plessey a 50 per cent cost sharing grant for production technology. In addition, the National Research Development Corporation awarded Marconi-Elliott, Ferranti and Plessey £12m. between 1969 and 1973 for R&D and production refinement in integrated circuits (Tilton, 1971). Finally, the RRE, SERL, the Post Office Research Station and other military institutions were granted £1.7m. (£4.2m.) for in-house R&D in 1968 (Golding, 1971, p. 347).

The French case was similar to the English case in that the French government was involved in a major restructuring of the domestic industry during the 1960s. It favoured the establishment of Sescosem, a merger between Sesco (Thomson) and Cosem (CSF). Between 1969 and 1973, Sescosem received 20m. frs. ($3.8m.) in R&D support.[96]

Several French programs supported R&D in semiconductors. The Plan Calcul of 1967–70 provided 91.6m. frs. ($18.7m.) for R&D in components (Dosi, 1981). The military followed an R&D and procurement policy that was directed toward silicon integrated circuits. As innovative as their objectives may have been, however, the size of military support remained small compared to the amounts given to firms by the American military. Moreover, these military funds were only channelled into large, established corporations. The French PTT and CNET also financed R&D in semiconductors, although to a limited degree. In 1968 27 per cent of the R&D in semiconductors in France was government financed (see Table 5.25).

Finally, in Italy there was no specific public policy towards the electronics industry during the 1960s. Some government funds (loans and R&D subsidies) were channelled into the manufacturing industry through the 'Applied Research Fund' of 1968.[97] However, by the early 1970s semiconductor producers had received very little of these funds.

6 The evolutions of the American and Japanese industries

6.1 *The American industry: the world leader*

During the 1960s the American semiconductor industry became world leader in both innovation and production, and far outpaced the European industry. The American industry showed a continuous record of both major and minor innovations, had increasing market shares in the European market, a surplus in the semiconductor trade balance and made direct foreign investments in Europe.

In particular, the American semiconductor industry became world leader in the digital integrated circuits market, the most technologically dynamic and the fastest growing market.

For the most part, American merchant producers were responsible for the performance of the American industry in the digital integrated circuit market. Texas Instruments and Fairchild, together with Motorola, became world leaders in the industry, as Table 5.26 shows. These two firms were joined by a large number of new merchant producers, who entered the American industry during the 1960s.[98] These producers believed that digital integrated circuits would prove the fastest and largest integrated circuit market and were strongly committed to digital integrated circuit production.

During the 1960s the major diversified vertically integrated receiving tube producers pulled out of the digital integrated circuit market.[99] Vertically integrated receiving tube firms continued to produce mainly discrete devices and linear integrated circuits. General Electric, for example, pulled out of mass production of digital integrated circuits and continued to produce discrete and power semiconductor devices, partly for internal consumption. Sylvania, following

Table 5.26 Firm shares of the semiconductor market and of semiconductor patents in the United States

Firm	Share of semiconductor patents in United States (percentage of total patents)			Market share in United States (percentages of total market)
	1961–4	1965–8	1961–8	1966
Bell-Western Electric	15.1	9.8	12.0	9
IBM	16.2	10.4	12.8	na
General Electric	10.5	13.0	11.9	8
RCA	9.3	12.1	10.9	7
Westinghouse	5.9	10.6	8.7	5
Texas Instruments	5.1	8.4	7.0	17
Motorola	2.4	5.3	4.1	12
Honeywell	5.6	2.0	3.5	na
Sperry Rand	2.7	3.6	3.3	na
General Motors	2.0	4.2	3.3	na
Hughes	2.4	3.1	2.8	na
Philco-Ford	3.1	2.1	2.5	3
Sylvania	2.7	2.0	2.3	na
Clevite	2.8	0.8	1.9	na
Bendix	1.2	2.0	1.7	na
ITT	2.4	1.3	1.7	na
Fairchild	1.0	1.7	1.5	13

Note: na = not available
Source: Tilton (1971).

its defeat in the logic war, also exited from integrated circuit production.[100] Westinghouse, on the other hand, remained involved in integrated circuits for some time;[101] it failed in the molecular electronics project supported by the Air Force in the early 1960s, however, and stopped production of digital integrated circuits later in the 1960s. Finally, RCA produced digital integrated circuits and was innovative in MOS technology but continued to focus its production on discrete devices and linear integrated circuits for consumer products.[102]

Western Electric was the only major American vertically integrated receiving tube producer that remained a major captive producer of semiconductor devices during the 1960s. In contrast to General Electric, RCA and Westinghouse, which were all diversified corporations, however, Western Electric was a specialized corporation. It therefore had well-defined uses for its semiconductor devices. This characteristic, coupled with the large size of its internal market and the support it was given by the military, allowed Western Electric to reap the benefit of economies of scale in semiconductor production.

The failure of established vertically integrated receiving tube producers in the digital integrated circuit market was therefore common to the United States and Europe. In both areas these corporations lacked rapid adaptation and full commitment to the new technology; they remained centred mainly on electronics final goods, and considered their production of semiconductors as dependent on these markets.

Similarly to the European vertically integrated receiving tube producers, American vertically integrated receiving tube producers were also unsuccessful in computer production. Sylvania, GE and RCA had to pull out of the computer industry during the 1960s and early 1970s (Katz and Phillips, 1982).[103] It is difficult to ascertain, however, whether the failure of digital integrated circuit production affected the competitiveness of the computer production of these firms or whether, more probably, the failure of computer production affected the competitiveness of digital integrated circuits production of these firms. It is only possible to claim that the lack of an internal market for digital integrated circuits deeply influenced the decision of these firms to discontinue their production of digital integrated circuits.[104]

The entry of a large number of merchant firms producing integrated circuits was the result of the extent of defence and space procurement. During the first half of the 1960s, the American Air Force used integrated circuits in its Minuteman Program and NASA used integrated circuits in its Apollo Program (Levin, 1982). This demand accounted for the total sales of integrated circuits in 1963-4 and most of the sales throughout the following year.[105] In the second half of the 1960s, however, the defence market in the United States decreased in relative importance: in 1965 the American military and NASA absorbed 28 per cent of the total production of semiconductors and 72 per cent of the total production of integrated circuits. In 1968 these percentages decreased to 25 and 37 per cent, respectively (see Table 5.27) (Levin, 1982, p. 60).

In addition, the growth of merchant firms producing integrated circuits of the digital type was a result of the extent and the high growth rate of the American computer market. As Table 5.28 shows, by the end of the 1960s, in fact, the computer market had surpassed the military, becoming the major American market for integrated circuits. This large and growing computer market stimulated innovative efforts and absorbed the production of digital integrated circuits of merchant firms.

Similiarly to the integrated circuit case, the initial growth of the computer industry in the United States was the result of substantial support through public procurement. The military, in fact, purchased a large share of the production of computers up until the mid-1960s (Katz and Phillips, 1982). In addition, public support helped new firms, such as CDC and DEC, enter the computer industry in the early-1960s. Finally, the government also supported

134   *The integrated circuit period*

Table 5.27 United States government purchases of semiconductors and integrated circuits, 1960–70

	Absolute value		As a percentage total output	
	Semiconductors ($ millions)	Integrated circuits ($ millions)	Semiconductors (%)	Integrated circuits (%)
1960	258		48	
1961	222		39	
1962	223	4	39	100
1963	211	15	35	94
1964	192	35	28	85
1965	267	57	28	72
1966	298	78	27	53
1967	303	98	27	43
1968	294	115	25	37
1969	247		17	
1970	275		21	

Sources: USDOC-BDSA in Levin (1982) for semiconductors and Tilton (1971), p. 63 for integrated circuits.

Table 5.28 Semiconductor end user markets in the United States (percentage of total market)

	1960	1968	1972
	%	%	%
Computers	30	35	28
Consumers	5	10	22
Military	50	35	24
Industrial	15	20	26
Total	100	100	100
Value ($ millions)	560	1211	1378

Source: Finan (1975) for the United States.

R&D in the industry: in 1965 the US government granted $300m. to R&D in computers (49 per cent of total R&D in the industry).[106]

The growth of the American computer industry was accompanied by a successful performance on world markets. The American competitive position in the European market, in fact improved during this period (OECD, 1969). Some of these American firms (i.e. IBM)[107] refused to license any European competitor. Other firms, however, developed extensive arrangements with European producers: RCA licensed several European firms (English Electric, AEG-Tele-funken, Siemens and Philips) and GE entered the European market by gaining control of both Bull and the computer production of Olivetti.

The size, growth and competitiveness of the American computer industry had another effect on the American semiconductor industry: it determined a new wave of upstream integration of computer producers into digital integrated circuit production. During the 1960s computer producers such as IBM, Burroughs, Honeywell, and Sperry Univac began producing custom-integrated circuits in-house.[108]

Contrary to the case of vertically integrated receiving tube producers, American computer producers were strongly committed to digital integrated circuit production. These types of circuits, in fact, were at the heart of computers, and therefore of their competitiveness.

Given their size, these computer producers, and particularly IBM, could afford to have a fully captive or at least a tapered integration of digital integrated circuits. In 1967 IBM, for example, had 19,773 computers installed in the United States and 3,145 installed in France, Britain and West Germany. In 1967 Honeywell had 1,800 computers installed in the United States and 140 in France Britain and West Germany; these figures were respectively 1,675 and 80 for Burroughs, and 4,778 and 369 for Sperry (OECD, 1969).[109] European computer producers, on the other hand, were much more limited in size. In 1967 ICL had 1,060 computers installed in these three countries, GE/Bull 907, Zuse 226, Siemens 129, Telefunken 26, and Philips-Electrologica 19 (OECD, 1969).

The captive production or the tapered integration of the largest American computer producers did not result in a reduction of the overall external demand for digital integrated circuits. For American merchant producers, in fact, the computer market grew steadily, and the entry of new, non-vertically integrated producers such as Control Data Corporation and Digital Equipment provided a large, external computer demand for semiconductor devices and integrated circuits.

Therefore, during the integrated circuit period, the evolution of the American semiconductor industry was characterized by a large-scale entry of new merchant producers and of computer producers into digital integrated circuit production, and by the large-scale departure of old merchant producers and old vertically integrated receiving tube producers from this production. As examined earlier, in several cases this was not an exit from semiconductor production but rather an exit from specific segments of the market—mainly the fastest growing or the most technologically advanced segments.

At a more general level, however, within the American industry the entry and exit of firms during the 1960s produced an automatic change in the routines and commitments to different technologies. These changes occurred through a selection of firms incorporating different routines, rather than through a selection of routines within established firms. While this second process was slow and characterized the European industry, the first process provoked a prompt reaction by the American industry to the new silicon-planar-(digital) integrated circuit technology.

6.2 The Japanese industry: the protection of the domestic market

During the integrated circuit period vertically integrated receiving tube producers continued to dominate the Japanese industry; they remained specialized in transistors for consumer applications and they were late to enter mass production of standard digital integrated circuits. Even though NEC produced limited quantities of integrated circuits in the early 1960s, only in 1968, with Texas Instruments' direct foreign investment in Japan, did local mass production of standard digital integrated circuits begin.[110]

The international performance of Japanese semiconductor firms deteriorated in the late 1960s. After 1967, in fact, Japan had a trade deficit in semiconductors, due to the large increase in imports of integrated circuits.

The specialization of the Japanese semiconductor industry in transistors and the delay in the mass production of digital integrated circuits can be explained by demand factors. Consumer demand continued to dominate the Japanese market during the 1960s. This type of demand mainly required transistors, and a limited amount of linear integrated circuits, but not digital integrated circuits. Demand for digital integrated circuits to be used in calculators only began in the late 1960s.[111] Sharp, for example, one of the major producers of calculators in Japan, used to purchase integrated circuits from the American firm Rockwell International. Only in the early 1970s did Sharp produce large quantities of semiconductor devices in-house.[112]

Demand for semiconductors coming from the computer market outpaced the demand from the consumer market only during the 1970s. During the 1960s the Japanese computer industry remained of limited dimensions and was constituted by six firms: NEC, Fujitsu, Hitachi, Mitsubishi, Toshiba and Oki.[113] In the early 1960s these firms did not have advanced technological capabilities in computers and had to purchase licences from American computer producers.[114] Only later in the 1960s did Japanese computer producers begin to develop internal technological capabilities in computers.

The government supported the Japanese computer industry during the 1960s in several ways. In the early and mid-1960s it fostered the creation of two R&D cartels among Japanese producers.[115] During the 1960s it induced the six Japanese computer producers to create the JECC (Japan Electronic Computer Company), a financial company with preferential access to sources of finance. More importantly, it protected the domestic market from import penetration with a 'Buy Japan' policy of procurement and with tariff barriers[116] and direct foreign investments through the Foreign Investment Law (1950) and the Foreign

Exchange and Trade Control Law (1949).[117] Only IBM had production facilities in Japan because it had been present in Japan before the Second World War.

The market share of the Japanese computer industry improved during the 1960s. The share of the value of the computer shipments of foreign producers declined in the early 1960s and then stabilized around 40 per cent for the rest of the decade, with IBM accounting for most of this share.[118]

During the 1960s the protection of the domestic market also played an important role in the semiconductor industry, by establishing high tariff barriers and by impeding direct foreign investments from American semiconductor producers. While several American firms opened facilities in Europe during the 1960s, only Texas Instruments was allowed to locate production facilities in Japan, in 1968. Texas Instruments, however, had to establish a joint venture with Sony, to hold only a 50 per cent share in this joint venture, to give integrated circuit licences to NEC, Hitachi, Mitsubishi, Toshiba and Sony, and to maintain a share of the Japanese market lower than 10 per cent (Tilton, 1971).

While the Japanese semiconductor industry had a technological and productive capability in (digital) integrated circuits during the 1960s which was not superior, and was eventually inferior, to the European one, it in fact enjoyed the protection of the domestic market at both the semiconductor and the computer level. This protection allowed the two world leaders alone—Texas Instruments and IBM—to be present on the Japanese market, but with well-defined limits and constraints imposed upon them. The protection of the home market at both the input stage and the final product stage was to play an important role in the evolution of the Japanese semiconductor industry during the LSI period.

Conclusions

During the 1950s and early-1960s a new technological regime was introduced in the semiconductor industry: the silicon-planar-integrated circuit regime. This regime radically transformed the existing semiconductor industry that was based on transistors. Semiconductor devices became sub-systems of electronic systems and the design of these devices became closely interrelated with the design of electronics final goods. As a result, electronics final-goods producers were given a new incentive to integrate upstream into the R&D and production of integrated circuits. In this regime, technological change also became more cumulative. Consequently, firms which were innovative in a specific product range in one period were likely to be innovative in that specific product range in the following period as well. It is important to note that the diffusion of integrated circuits among users and suppliers was a dynamic phenomenon. The diffusion of integrated circuits among users did not simply involve the substitution of transistors with integrated circuits in existing applications, but also opened up entirely new applications. The introduction of the integrated circuit, therefore, set in motion an expansion of demand for integrated circuits.

The diffusion of the integrated circuit among suppliers, on the other hand, was not a smooth and immediate process. Both in the European, as well as in the American and Japanese industries, several firms continued to produce discrete semiconductor devices and to disregard the integrated circuit in the belief that it was not an innovation which was going to have a major impact on the industry. In particular, firms previously successful in the old technological regime—such as Philips and Siemens in Europe and General Electric and RCA in the United States—did not rapidly adopt the new technology. The vested interests and the success of these firms in the then-current technological regime led them to disregard the importance of the new technological regime.

Therefore, while at a general level late entrants are at a disadvantage compared to incumbents, in the case of appropriable technology and cumulative technological change late entrants may have an advantage over incumbents as regards radical changes and switches in technological regimes. In fact, in the semiconductor industry, while established and successful producers did not have any major advantage over late entrants in the germanium-alloy-discrete device technological regime or during the switch from this regime to the silicon-planar-integrated circuit one, they had an advantage over late entrants once this latter regime had been established.

In addition, on a more general level, the way in which firms switched to the new technology took quite different forms in the United States and in Europe. In the American industry the change in routines resulted from a change in firms: a selection process occurred through which firms which had the 'fittest' routines survived and prospered. In the European industry, by contrast, the change in routines occurred within firms rather than through a change in firms.

The slow transfer to silicon-planar-integrated circuit technology and the lack of commitment to large-scale production of digital integrated circuits of the European vertically integrated producers was the result of the specific linkages and interdependencies of the European semiconductor market. The European demand structure was in fact characterized by a relatively high importance of consumer and industrial demand which privileged discrete devices first, and linear integrated circuits later. In addition, vertical integration of these European producers in the final market reinforced the commitment to discrete devices and linear integrated circuits, but not to digital integrated circuits. It is enough to name the final products produced by the major European producers for an understanding of this mechanism: Philips and AEG-Telefunken produced electronics consumer goods, Siemens, Thomson-CSF and Plessey produced telecommunication and industrial equipment. And, as a counter-example, it is indicative that the two major European success stories of the 1960s had for major actors a firm (SGS) which chose an American partner (Fairchild) and which was not *de facto* a vertically integrated producer,[119] and a vertically integrated producer (Ferranti) with the right technological beliefs and an initial commitment to the computer market.

An analysis of their histories serves to emphasize, however, that commercial success in the long run requires that correct technological choices must be complemented by the development of an independent technological capability and by a market ready to absorb the new products. In the case of SGS and Ferranti the lack of these factors provoked the rapid decline of these firms.

While Japan had demand and supply factors very similar to those of Europe, the United States did not. The linkages and interdependencies of the American market determined a commitment of American producers to, and innovations in, silicon-planar-integrated circuits first, and digital integrated circuits later. In the United States the military and computer markets 'pulled' semiconductor technology in specific directions: the military market pulled it towards the silicon integrated circuit while the computer market pulled it towards the digital integrated circuit. Because of their large size and high growth rates, several American computer producers, such as IBM, vertically integrated upstream into digital integrated circuit R&D and production. Contrary to other semiconductor producers vertically integrated in different final markets (such as consumer and industrial markets), their specific electronic final market (computers) led them to a commitment to R&D and production in digital devices.

During the integrated circuit period, the European case strongly differed from the American and the Japanese cases as far as the public policy factor is concerned. While American government policy (the military and NASA) strongly supported the development of integrated circuits through public procurement and R&D support, and Japanese government policy protected the domestic market through tariff barriers and foreign direct investments blocks, European government policy neither strongly supported the domestic semiconductor industry nor protected the domestic market.

Therefore, by the beginning of the LSI period in the early 1970s, American merchant producers dominated the world digital integrated circuit market— the most technologically dynamic and fastest growing segment of the semiconductor market. These merchant producers had been able to enter the European market through exports and direct investments and to compel many European producers to retreat from it.

Notes

1. A layer of silicon dioxide was created on the surface of a silicon substrate. Then, photolithography was used to etch holes in the oxide layer which acted as a mask for the diffusion. Impurities were then diffused into the substrate. Several diffusions could be carried out serially.
2. Devices in these groups could be produced with either a bipolar or an MOS technolgy. See Chapter 2.

3. The following discussion is based on Young (1979).
4. The input comes through a resistor into the base of an inverting transistor. RTL has low-power dissipation.
5. The input comes through a diode and the output comes from the collector of an inverting transistor. DTL has a slow speed.
6. A development of the DTL is the integrated injection logic (I2L) which is a very compact, high-density and high-speed circuit.
7. In the ECL circuits a pair of transistors is coupled by their emitters. ECL circuits are the fastest type of logic circuits.
8. The input comes through a transistor. TTL is used in high-speed application and has medium-power dissipation. Sylvania's TTL was called SUHL.
9. The first types of MOS ICs were positive channel MOS and negative channel MOS. Positive channel MOS have a positive type of semiconductor as the conducting channel. In 1969 the complementary MOS (CMOS) logic circuit was introduced by RCA. MOS integrated circuits consume little power and are very resistant to stray noise (pulses). However, they are also slow and have a low packing density. They have been used extensively in digital watches. In 1973, RCA introduced the Silicon on Sapphire (SOS), which was aimed at increasing the speed of MOS circuits. In this device, the sapphire substrate acts as an insulator to increase the speed of the circuit operation. SOS/MOS are very expensive. In 1969, Fairchild introduced the charge-coupled device (CCD), which was composed of an array of MOS capacitors. CCD can be used for signal processing, both as an analog or a digital serial memory and as a dynamic filter. It can also be used in sold-state cameras (Young, 1979; Wilson, Ashton and Egan, 1980).
10. Later in the 1960s Fairchild introduced several successful types of linear integrated circuits. In 1964 Fairchild introduced the 702 linear integrated circuit and later the 709. The 709 had a large commercial success and was based on a different design than the 702. Instead of reproducing the former circuits composed of discrete elements on the semiconductor wafer, the 709 used a minimum amount of resistors and capacitors. This decreased the dependency of the circuit on resistors and capacitors (Dummer, 1978, p. 147).
11. Epitaxy was introduced by Bell in 1950. The epitaxy method represented an improvement in the planar process. The epitaxial transistor was able to handle higher frequencies and had a higher power capability.
12. For an explanation of these process innovations, see Dummer (1978), Young (1979), United States Department of Commerce (1979), Wilson, Ashton and Egan (1980), Braun and MacDonald (1982).
13. The changes in the characteristics of technological development in the semiconductor industry which occurred between the transistor and the integrated circuit period demonstrate the differences between a 'science-based' and a 'cumulative' technology, discussed in Nelson and Winter (1982).
14. As examined above, large teams of technicians (rather than single individuals) using complex equipment played an increasing role in the innovative process. For a description of the working of R&D organizations, see Allison (1969).

15. See Nelson (1982) for a general discussion and Braun and MacDonald (1982) and Levin (1982) on technological advances based on engineering.
16. From a graph of Tilton (1971), p. 124.
17. In Germany, for example, the ratio of domestic production to the total consumption of semiconductors decreased from 0.8 in 1965 to 0.5 in 1973 (Scholz, 1974, p. 126). This decrease in self-sufficiency was mainly due to the increasing trade deficit in integrated circuits; the trade deficit in discrete semiconductors in terms of the overall German market remained below 10 per cent until 1967 and below 20 per cent during the period 1967–72 (Bundes Ministerium für Forschung und Technolgie—BMFT—1974). However, figures related to international trade in semiconductors in the late 1960s/early 1970s should take into account trade to and from American and European offshore assembly plants. For example, the negative trade balance in integrated circuits in the German market jumped from approximately 25 per cent (1971) to approximately 65 per cent (1973) of the German integrated circuit market (BMF, 1974). However, the share of imports would be lower if the trade to and from offshore asembly plants were taken into account (Scholz, 1974, p. 126).
18. At a general level, these two situations are similar respectively to a science-based industry and to a cumulative-technology industry represented in the Nelson and Winter (1982) models. The disadvantage of a late entry during the integrated circuit period is related to specific product families, such as bipolar integrated circuits, MOS integrated circuits, etc. Wilson, Ashton and Egan (1980) analyse some major technological races in the American industry and reach the conclusion that being first is not necessarily an advantage. Their conclusion applies to innovative American firms with a relatively short period of delay in entry.
19. Source: interviews. In 1965 Philips merged its Electronica division, which produced active components (electron tubes, cathode ray tubes and semi-conductors) with its Icoma division, which produced passive components. The new division was called Elcoma.
20. Scholz (1974), p. 139; and interviews.
21. Plessey was the first firm to develop a model of an integrated circuit (in 1957) from a contract with the RRE. Source: OECD (1968), p. 61.
22. Scholz (1974); and interviews.
23. Finan (1975) has a detailed analysis of this phenomenon.
24. See Caves (1971) for an analysis of externalities and market imperfections as causes of direct foreign investments and Rugman (1981) for an analysis closer to the present one.
25. This argument is closely related to the transaction costs explanation of vertical integration that was discussed earlier. Both arguments stress the importance of the market for information and knowledge. However, in the analysis of firm-specific organization and vertical integration, the focus was on the integration of different functions, activities or production processes. In the present analysis of the horizontal types of direct foreign investments, the focus is on the reproduction of the same advantages of a firm in a different location. Teece (1981) argues that because of the

presence of transaction costs, the multinational enterprise is the most efficient means of transferring technology.

26. The first of these factors is related to firm size; the larger the firm, the less costly the foreign investment. Large firms have more financial resources than small firms so that the 'big jump' of direct foreign investments is less burdensome. In addition, large innovative firms rely on a large pool of skilled workers and technicians. Often the demand for skilled workers will be in excess of the supply in the home country so that large firms will find it necessary to locate activities in foreign countries. In addition, direct foreign investments may be made in order to protect a firm's market share in foreign countries. These market shares may be threatened when firms in the same strategic group make investments in the same country. See Knickerbocker (1973) for an analysis of the clustering of direct foreign investments at the industry level.

27. According to Dunning (1977, 1979), the analysis of direct foreign investments must be based on the distinction between location-specific endowments of countries and the ownership-specific endowments of enterprises. In addition, Hymer (1976) and Kindleberger (1969) argue that multinational enterprises must have some firm-specific advantages in order to compete with domestic firms. This point is reaffirmed by Rugman (1981).

28. There existed no specific pattern of licensing in the semiconductor industry. First, as a result of the liberal policy which Bell followed and as a result of the dispute between Texas Instruments and Fairchild over the integrated circuit, basic patents became available to competing firms. Second, small innovative firms followed a second source policy (Finan, 1975). Third, Texas Instruments only fully licensed its own subsidiaries and Fairchild established a joint venture in Europe (SGS). Fourth, European vertically integrated receiving tube producers obtained licences for integrated circuits, mainly from American vertically integrated receiving tube producers: only later in the 1960s did Fairchild begin granting them licences.

29. Texas Instruments claimed that there was no tehcnology gap between the United States and Europe as far as uses were concerned; *Business Week*, 1 March 1970.

30. The other two partners in SGS-Fairchild were Olivetti and Telettra, each with one-third of the total capital.

31. Between 1957 and 1961 SGS had lost L.686m. ($1.1m.).

32. For a detailed anlaysis of the offshore assembly, see UNCTAD (1975) and Ernst (1981).

33. In 1971 there were approximately 17,000 employees in electronics components manufacturing and 10,000 in semiconductor manufacturing in American firms; 5,000 and 4,000 respectively in European firms. These figures were also small for Japan. However, by 1974 the number of employees in Japanese controlled firms located in less developed countries was higher than the number of employees in European controlled firms. Source: UNCTAD (1975), pp. 17–18.

34. Truel (1980) claims that an alternative to offshore assembly was the complete automation in the home country of the assembly operations.

Automation was chosen by Japanese semiconductor producers in the first instance. Later during the 1970s, however, all countries began to automate their assembly operations and, as a result, the incentives to establish or to continue offshore assembly operations diminished. See Ernst (1981).

35. The term 'logic war' has been used by Wilson, Ashton and Egan (1980).
36. Sylvania's SUHL-TTL had faster switching speeds than the F930 DTL.
37. In the American market, the wholesale price index for a high-speed TTL integrated circuit went from 100 (1968) to 37.3 (1971). Webbink (1977), p. 77.
38. *Business Week*, 22 August 1970.
39. Source: interviews.
40. Philips bought Electrologica in 1960.
41. However, GEC soon began buying these circuits from Texas Instruments. Source: interviews.
42. Source: interviews.
43. Source: interviews.
44. Siemens underwent some major structural and organizational changes during the second half of the 1960s. In 1967 it purchased Zuse, a computer producer, and in the same year the Bosch-Siemens Hausgerate (consumer goods) was established. In 1969, a new organizational structure, with six major branches, was established; see Siemens Berichte, various years.
45. The DTLZ circuits were used in-house in machine tools.
46. The commercial production of the ECL never materialized, however; only a symmetrical ECL (SECL) which was developed for use in computers was ever utilized in consumer devices.
47. Source: interviews.
48. During the 1960s the fourth largest German semiconductor producer, Intermetall, experienced a change in ownership and consequently a change in productive specialization. In 1965 ITT purchased Intermetall from Clevite and joined it to an already existing small ITT transistor producer in Germany.
49. Significantly, in the late 1960s *Electronics* magazine entitled a report on the German semiconductor industry 'Germany, where the linear IC reigns' (*Electronics*, 15 September 1969).
50. The history of the British firms is based on Golding (1971), Sciberras (1977) and interviews.
51. In 1962 GEC established ASM, a joint venture with Mullard (Philips): one-third of the capital was owned by GEC and two-thirds by Mullard. In 1969 GEC retreated from the ASM venture after having purchased the semiconductor operations of AEI and EE in 1967.
52. Golding (1971), Sciberras (1977) and interviews.
53. As examined earlier, some major structural changes occurred in the French electronics industry during the 1960s. In 1966 Thomson-Houston merged with Hotchkiss-Brandt, thus becoming Thomson-Brandt. In 1968, Thomson-Brandt took control of CSF, and formed the Thomson-CSF company. Thomson-CSF became the major French producer of industrial and

professional electronics products in France. At the semiconductor level, the creation of Thomson-CSF resulted in the creation of Sescosem in 1969. Sescosem was the result of a merger between Cosem (CSF) and Sesco (Thomson-Brandt): Sesco was a joint venture between Thomson-Houston (51%) and General Electric USA (49%).

54. A relevant part of CSF's market was composed of the Defence, the PTT, and other types of public procurement (*Expansion*, April 1968).

55. See Pottier-Touati (1981).

56. As previously seen, SGS-Fairchild was a joint venture between the Italian SGS (owned by Olivetti and Telettra) and Fairchild.

57. In Italy another domestic semiconductor producer existed: ATES. ATES was established by SIT-Siemens in 1961. It produced germanium diodes and transistors for the Italian and German consumer markets. During the 1960s, because of its technological link with Siemens, ATES manufactured germanium mesa transistors.

58. Olivetti controlled one-third of the stock of SGS-Fairchild and absorbed approximately 40 per cent of SGS's production. Source: interviews.

59. Source: interviews.

60. In 1969, for example, sales of digital integrated circuits constituted more than one-fifth of SGS's sales, while sales for the computer market were more than one-quarter of SGS's sales. Both values were higher than for the average European semiconductor producers in that year. Sources: SGS-ATES and interviews.

61. In fact, Olivetti bought both Fairchild's and Telettra's shares in SGS in 1968/9.

62. Source: interviews. In 1969 Fairchild opened a production facility in Wiesbaden (Germany).

63. Flaherty (1981) argues that discussion about control should be kept separate from discussion about financial integration. She demonstrates that financial integration occurs in the case of (quantity) control when the relationships between input producers and users is long-term and negotiation between them is continuous. Her approach is very close to the discussion in this section. However, in her discussion of (quantity) control she assumes a fixed proportions production function of the downstream unit. As a consequence, uncertainty in the production function related to technological change is not taken into account.

64. For a general discussion, see Pelc (1980).

65. These licences, of course, would be for integrated circuits that were slightly modified from the original design.

66. It should be noted that during this period some corporations vertically disintegrated. These firms absorbed limited quantities of custom and standard integrated circuits and did not profit from sales of semiconductors on the external market. In addition, these firms believed that the in-house production of integrated circuits would not affect their competitiveness and innovativeness in final products. They therefore moved out of the production of integrated circuits, remaining only in the production of discrete devices. This is true of CGE (France) in 1967, GEC and AEI

(Britain) respectively in 1962 and 1968, and General Electric and Westinghouse (USA) during the 1960s.

67. Such as Grundig (consumer electronics) or ICL (computers).
68. As will be examined later, in the United States a wave of upstream vertical integration by computers as well as other specialized electronics final goods producers occurred.
69. This concept has been used by Scriberras (1977) to explain the survival of British firms during the 1960s and 1970s.
70. Source: interviews.
71. Source: interviews.
72. In France, Thomson-Houston, Hotchkiss-Brandt and CSF produced a diversified range of final products for both industrial and consumer electronics. Following the merger Thomson-Houston with Hotchkiss-Brandt and the takeover of CSF in 1968, the new subsidiaries Thomson-Brandt and Thomson-CSF produced, respectively, consumer electronics products and a whole range of electronic industrial and telecommunications equipment.
73. Sciberras (1977) claims that in the early 1970s 60 per cent of Lucas' semiconductor turnover consisted of in-house sales (Sciberras, 1977, p. 192).
74. Plessey was the first firm in the world to fabricate a linear integrated circuit; in 1968 linear integrated circuits accounted for 80 per cent of Plessey's semiconductor sales (Golding, 1971).
75. This category includes electrical and electronic components.
76. Source: interviews.
77. Public procurement and the military absorbed a large share of professional electronics equipment, such as radars and telecommunication equipment. In France, for example, in certain types of electronics products the public adminstration and the military constituted 80 per cent of the domestic demand (France-V Plan, 1966).
78. See also *Business Week*, 19 February 1966.
79. In addition, in 1967 Siemens acquired 70 per cent of the capital of Zuse, a major German computer manufacturer which had previously been owned by Brown Boveri. Siemens made a major effort in computer manufacturing in the second half of the 1960s. According to Harman (1971) Siemens spent $125m. on computers in 1966. In the German market, Siemens ranked second only to IBM.
80. See Rösner (1978), p. 61.
81. See OECD (1969) and (1977).
82. OECD (1977), Hu (1973) and Jublin-Quatrepoint (1976) provide a detailed account of the take-over.
83. The choice to produce mechanical and electromechanical rather than electronic office equipment products was eventually acknowledged to be a major mistake by Olivetti (*Business Week*, 26 October 1974).
84. The Programma 101 used transistors in its first version and integrated circuits in its second version. Through this production, a collaboration between SGS and Olivetti began.

85. See Harman (1971), p. 32.
86. Dosi (1971), Golding (1971) and Grant and Shaw (1979).
87. Source: interviews.
88. Source: interviews.
89. Even though Olivetti adopted the TTL 9000, produced by SGS, in its peripheral and numerical control machines. Souce: interviews.
90. Several sectors (biology, transport, raw materials and electronics) were supported. Federal Republic of Germany, BMFT (1980).
91. From 1969 to 1970, opto-electronic components were given DM0.3m. ($76,923) and basic developments and new components DM1.3m. ($333,000). Federal Republic of Germany, BMFT (1974), p. 37.
92. A non-profit research institution with the task of basic research in electronics.
93. A military committee (see Chapter 4).
94. During the late 1950s and early 1960s a thin film hybrid programme worth £4m., sponsored by the RRE and CVD, failed. Mullard received the main contract but it withdrew from the programme in 1967 because the integrated circuit had displaced thin film hybrid circuits (Golding, 1971, p. 354).
95. Golding (1971), p. 361. Ferranti received the grants for the development of integrated circuit functions; Elliott Automation received grants for work on beam lead and MOS technology. Source: interviews.
96. Pottier and Touati (1981).
97. The fund was originally endowed with L.100b. ($160.5m.) to support applied research in all sectors of the economy. It was managed by IMI (a public financial institute).
98. In the history of the American semiconductor industry, there have been three peak periods of entry: 1952–3, 1959–63 and 1968–73. See Wilson, Ashton and Egan (1980).
99. Kraus (1973); Braun and MacDonald (1982).
100. This occurred in 1970. See Wilson, Ashton and Egan (1980).
101. Westinghouse, a vertically integrated receiving tube producer, had an approach to the integrated circuit based on germanium and on a complex manufacturing process which was not suited to mass production (OECD, 1969).
102. RCA was the first to introduce CMOS technology. See Wilson, Ashton and Egan (1980).
103. RCA produced IBM compatible computers, while General Electric tried to develop its own technology (Brock, 1975). Among the reasons for their failure are the consumer electronics vocation of RCA and the inadequate allocation of resources by General Electric to computer production (Katz and Phillips, 1982). General Electric sold its operations to Honeywell and RCA sold on to Sperry Rand. General Electric's and RCA's failure in computers was similar to the failure of these firms in digital integrated circuit production.
104. The situation of General Electric and RCA was similar to that of Siemens during the 1960s; they were diversified vertically integrated corporations

with an unsuccessful computer production. However, while General Electric and RCA were eventually forced to exit from computer production, Siemens remained in the market partly for strategic reasons (Von Weiher and Goetzeler, 1984) and partly because of the support it received from the German government.

105. Levin (1982). American military demand played a major role in the introduction of the silicon transistor and had a major, though indirect, influence on the introduction of the integrated circuit. Texas Instruments was given a particularly large contract in 1962 (OECD, 1968, p. 63).

106. OECD (1969), p. 136. This figure may be compared to the $4m. (16 per cent of total R&D) supplied by the French government to domestic producers for R&D in computers.

107. During the 1960s IBM increased its position of leadership with the introduction of the 360 series which was based on hybrid circuits (see OECD, 1969).

108. Source: Klesken and Dataquest. It should be noted that many of these firms produced custom-integrated circuits in very small quantities.

109. Hewlett Packard also began the captive production of a limited amount of semiconductor devices during the integrated circuit period. Hewlett Packard, however, was not specialized in computers, as IBM, Honeywell, Buroughs and Sperry were.

110. Borrus, Millstein and Zysman (1982); Nomura (1980).

111. In Japan the first electronic calculator was introduced by Sharp in 1964 (Nomura, 1980).

112. Masuda and Steinmuller (1981).

113. Matsushita was part of this group, but exited early from computer production.

114. Hitachi obtained licences from RCA, Mitsubishi from TRW, NEC from Honeywell, Oki from Sperry Rand, and Toshiba from General Electric.

115. The first R&D cartel obtained approximately $1m. in subsidies; the second one obtained approximately $40m. in subsidies. However, direct R&D support by the government remained limited.

116. The tariffs were 15 per cent on the central processing units and 25 per cent on peripherals.

117. Only Sperry Rand was allowed to open a plant in Japan, but it had to accept a minority share in a joint venture with Oki.

118. Borrus, Millstein and Zysman (1982); Nomura (1980).

119. Even though the difference in demand structure in Europe and the United States strongly affected the R&D and production strategies of the two firms. As seen previously, SGS-Fairchild had serious problems in trying to reconcile the American-driven focus of Fairchild on standard devices for the computer market and the European-driven focus of SGS on speciality and custom devices for the consumer market. In fact, these differences in focus represented one of the factors that eventually led to the end of the joint venture in 1968.

6 The large-scale integration period: lagging competitiveness and the restructuring of the European industry

In this chapter the evolution of the European semiconductor industry during the large-scale integration (LSI) period will be analysed. The LSI period covers the years between the introduction of the microprocessor (1971) and the early 1980s.

The chapter is organised in the following way: a discussion of the technological regime, the major innovations and the technological trajectory is presented in Section 1. The history of the evolution of the European semiconductor industry during the LSI period is contained in Section 2. The major factors affecting the evolution of the European industry (supply, demand and public policy) are discussed in Sections 3, 4 and 5. Finally, the evolutions of the American and the Japanese industries are analysed in Section 6 as a comparison with the European industry.

1 Technological regime, major innovations and technological trajectory

The LSI period begins with the introduction of the microprocessor by Intel and with the widespread use of memory devices. Microprocessors and memories led to the introduction of the microcomputer by Texas Instruments in 1972. Microprocessors, memory devices and microcomputers had an enormous impact not only on the semiconductor industry, but also on the electronics industry as a whole.

Microprocessors, memory devices and microcomputers established a new technological regime: the large-scale integration (LSI) regime. As previously mentioned, the term LSI refers to semiconductor devices with between 100 and 100,000 gates.

The new LSI technological regime was not something new from a technological point of view, because it represented a continuation of the former integrated circuit regime. It increased and opened up, however, completely new fields of applications for semiconductor devices. In addition, it enlarged the types of functions that a semiconductor device could perform: sub-system and system functions, rather than simple logic unit functions, as in the previous integrated circuit regime. The microcomputer, for example, is composed of a microprocessor, a memory and input-output devices. Table 6.1 provides an overview of the market for microprocessors in the United States.

The LSI technological regime was also based on the increasing digitalization of linear functions. Digital devices were more reliable and precise than linear

Table 6.1 Microprocessor markets in the United States, 1979 (percentage of total microprocessor market)

	4-bit	8-bit Single chip	8-bit Multi chip	16-bit Multi chip
	%	%	%	%
Consumer	81	25	10	0
Industrial	7	31	30	30
Computer	12	42	45	65
Defence	–	2	15	5
Total	100	100	100	100
Number of units	50m.	10m.	14m.	90,000

Source: Dataquest in *Mondo Economico* (23–30 August, 1980).

devices. They increasingly substituted linear devices in consumer electronics (digital audio and digital TV) and in telecommunication and industrial equipment.

Finally, the LSI technological regime was based on increasing interrelationships between semiconductor devices and computers and on increasing technological interdependencies among computers, electronic consumer goods, and telecommunications equipment. Computers and semiconductor devices, in fact, became closely linked technologically, because solid-state memories and single chip microprocessors formed the core of computers and initiated the development of new computer families (e.g. the minicomputer and the personal computer). The technological interdependencies among the computer, electronics consumer goods and telecommunication equipment increased because of the extensive use of microprocessors and the digitalization of linear functions in these products.[1]

The major innovations of the LSI period included product innovations such as the bipolar junction field effect and the bipolar-MOS combination,[2] the integrated injection logic (I2L),[3] and the vertical MOS (VMOS),[4] and several process innovations connected with LSI devices.[5] Major changes were made, for example, in photolithographic techniques in order to produce very large-scale integrated (VLSI) devices with micron or submicron channel widths and in order to develop lithographic methods that were more accurate than optical lithography.[6]

As during the integrated circuit period, technological change during the LSI period was cumulative. At the general technological level this meant that new technological advances were based on the advances that had been made previously. The microprocessor, for example, was based on advances in computer architecture, semiconductor memories, MOS technology and logic design; new memory devices were based on previous types of memory devices.

Incremental innovations and improvements gained increasing importance.

Opto-electronic devices, for example (i.e. light-emitting diodes, photodetectors, couplers, solar cells), underwent several improvements. In addition, high-power devices improved their speed and their capacity to handle power current and voltage, microwave devices were made that were able to handle both microwave and ultra-high frequencies, and analog-to-digital and digital-to-analog converters (interfaces) augmented their capability and were increasingly used in electronics equipment.[7]

At the firm level cumulativeness was associated with high appropriability: firms had high probabilities of retaining technological advantages in specific product/process ranges from one period to another. Intel, for example, was an innovator in microprocessors (the leader in 4-bit and 8-bit microprocessors, and among the leaders in the 16-bit) and in MOSRAM memories (the 1K MOSRAM, 4K MOSRAM and 16K MOSRAM). Mostek was among the leaders in the 4K MOSRAM and the 16K MOSRAM (Wilson, Ashton and Egan, 1980), while Fairchild was the innovator of the bipolar RAM and the 4K 12L RAM (Dummer, 1978).

During the LSI period semiconductor technology followed a trajectory toward miniaturization and toward the integration of hitherto separable functions. Miniaturization implied that the channel width of semiconductor devices would be reduced to micron or submicron size, while integration implied that semiconductor devices would be able to perform sub-system and system functions, rather than just simple logic unit functions. In addition, digital LSI devices, rather than linear devices, were increasingly used in electronics products and equipment.

LSI products and processes became increasingly complex. Design played an increasingly important role and required a capability in computer and logic architecture and in the use of the computer (CAD). LSI equipment also became increasingly sophisticated.[8]

During the LSI period new materials (other than silicon) expanded the performance and application of semiconductor devices, although the use of some of these new materials was still in an experimental stage. Gallium arsenide transistors, for example, were faster than silicon transistors and began to be used in microwave amplifiers and high-speed devices.[9] Moreover, solid-state lasers and optical fibres were used in devices for telecommunication equipment, optical devices were used for memories in electronics consumer goods,[10] and magnetic bubbles were used in memories.[11]

During the LSI period MOS technology replaced bipolar technology as the most widely used technology in digital devices (see Table 6.2). As previously seen, MOS integrated circuits consumed less power and had higher density capabilities than bipolar integrated circuits; these properties were particularly important in memory devices and, later, in microprocessors. By the early 1980s, in fact, several different types of MOS-integrated circuits had been developed: the P-MOS, used mainly in the mid-1970s, the N-MOS, widely used

Table 6.2 Semiconductor consumption

A. Integrated circuit consumption over total semiconductor consumption in the United States, Europe and Japan

	1974	1978	1983
	%	%	%
USA	56.0	68.4	83.4
Europe	39.0	52.8	66.7
Japan	50.6	55.2	66.8

B. Shares of semiconductor types in total consumption

	1979		
	USA	Europe	Japan
	%	%	%
Discrete	23.5 (1)	43.0	36.3
Linear IC	11.7 (2)	14.0	15.3
Bipolar IC	22.5 (3)	16.0	12.1
MOS IC	39.2 (4)	27.0	29.9
Other	3.1	na	6.4

(1) of which 3% to computer market
    and 97% to other markets in 1978 (C)
(2) of which 11% to computer market
    and 89% to other markets in 1978 (C)
(3) of which 37% to computer market
    and 63% to other markets in 1978 (C)
(4) of which 51% to computer market
    and 49% to other markets in 1978 (C)

Sources: A. *Electronics* (various years).
        B. SIA in United Nations (1983) p. 24.
        C. US International Trade Commission (1979) p. 23.

in the late-1970s–early-1980s, and the CMOS, which are fast becoming the most widely used of the MOS circuits. CMOS devices consume less power and perform better in many circumstances than the other types of MOS integrated circuits.[12]

2 The evolution of the European industry: a case of failure to catch up

2.1 *The world industry*

During the LSI period the world semiconductor industry continued to grow at a considerable pace. Between 1973 and 1982 total demand for semiconductors increased by 11.8 per cent annually. The demand for integrated circuits increased by 15.7 per cent, while the demand for discrete devices increased by 7.3 per cent. Within integrated circuits, the demand for MOS integrated circuits increased by 22.9 per cent, while the demand for bipolar integrated circuits increased by 8.6 per cent.[13]

During the 1970s and early 1980s the price of LSI devices continued to decrease. The price of a 16K RAM, for example, decreased by 90 per cent between 1980 and 1981 and the price of a 64K RAM decreased from $100 (1980) to $10 (1981) to $5 (mid-1982) (*The Economist*, 20 March 1982).

During the LSI period two semiconductor recessions occurred: one in 1974-5 and one in 1981-2. In 1975 in the United States, for example, the value of the total shipments of transistors and integrated circuits decreased, respectively, by 25 per cent and 19 per cent.[14] In Germany, the value of the production of discrete semiconductors and integrated circuits decreased, respectively, by 36 per cent and 6 per cent,[15] while in France these values decreased, respectively, by 25 and 30 per cent.[16] The 1974-5 recession induced American merchant producers to reduce their investments in additional plants and equipment. As a result, when the recovery took place in 1976, American firms were unable to keep up with demand, particularly in the fast-growing memory market. This demand was therefore satisfied by Japanese producers, who had not limited their investments during the recession. In the semiconductor recession of 1981-2 the sales of integrated circuits in Europe decreased significantly: total sales decreased from $2.3bn. in 1980 to $1.7bn. in 1981 (a decrease of 26 per cent) and stagnated in 1982.[17] The sales of memory devices, moreover, decreased by 22 per cent in 1981 on the European market (*The Economist*, 20 March 1982). This decline in sales on the European market affected American producers much more than European producers. During the 1981-2 semiconductor recession, however, American merchant producers did not restrict their investments in additional plants and equipment as they had done during the 1974-5 recession; they had learned a lesson from their experience in the previous recession.

During the LSI period the American semiconductor industry continued to be the world leader at both the innovative and the commercial level, but in the early 1980s it had to share its leadership with the Japanese industry in some technological areas. During the 1970s and early 1980s the United States continued to produce more than half of the total world production of semiconductors, and it was the largest market for semiconductors. Japan produced the second largest number of semiconductors while Europe was the second largest consumer of semiconductors until the late 1970s, when Japan took its place (see Tables 6.3 and 6.4).

The leadership of the American industry appears even greater when integrated circuit production is taken into account: the United States was the largest, and Japan the second largest producer of integrated circuits in the world, as Tables 6.5, 6.6 and 6.7 show.

Table 6.3 Production and consumption of semiconductors by major world area (percentage of world total)

	Unites States		Europe		Japan	
	Pro-duction	Con-sumption	Pro-duction	Con-sumption	Pro-duction	Con-sumption
	%	%	%	%	%	%
1970	57	na	16	na	27	na
1975	64	41[a]	17	31	19	23
1978	59	39[a]	13	26	28	27
1982	63	53	10	18	26	21

Notes: [a]with Canada
na = not available
The rest of the world must be included for the world total.

Source: Dataquest (1979, 1980) in Braun and MacDonald (1982), pp. 152, 153; for 1970, 1975, 1978 'Integrated Circuit Engineering' in UN (1983), p. 21, and *Elettronica* (10, 1984) for 1982.

Table 6.4 Production of semiconductors by country (United States = 100)

	1972	1978
United States	100	100
Japan	37	41
West Germany	9	7
Great Britain	9	4
France	6	7
Italy	3	2
Netherlands	7	2
Europe Total	35	24

Note: The production of American subsidiaries is included in the production of the host country.

Source: US Department of Commerce (1979) for 1972 and UN (1983) for 1978 (pp. 34, 47).

The American (and Japanese) industry also had a large share of the world market in LSI (and VLSI) devices. In the microprocessor-microcomputer market (4-, 8- and 16-bit), the American industry dominated in most areas: Intel (US) dominated the 4- and the 8-bit markets, and Intel, Motorola, Zilog, Texas Instruments and National Semiconductor (all American firms) dominated the 16-bit market.[18] Most European firms only produced devices as second source: Siemens for Intel devices, Sescosem (Thomson-CSF) for Motorola devices, SGS-ATES for Zilog devices.[19]

In the memory market, as well, the American and Japanese industries dominated the scene. The American industry produced most of the 1k dynamic

Table 6.5

A. Production of Integrated Circuits by major area
   (percentage of world production)

	1978	1982	1984 (forecast)
	%	%	%
United States	68	70	67
Japan	18	23	27
Western Europe	6	6	5
Others	8	1	1
Total	100	100	100
World Total	$7,012m.	$13,380m.	$19,140m.

B. Total consumption supplied by American firms, 1975
   (percentage of total consumption)

	All semiconductors	Integrated circuits
	%	%
Western Europe	58	78
Japan	11	20
United States	97	98

Sources: (A) ICE and estimates in Borrus, Millstein and Zysman (1982) for 1978 and UN (1983) p. 34 for 1982 and 1984. (B) WEMA in United States Department of Commerce (1979), p. 90.

Table 6.6 Production of integrated circuits by country, 1974 and 1980 ($ millions

	1974	1980
West Germany	112	306
Great Britain	100	153
France	74	111
Italy	29	77

Source: US International Trade Commission (1979), p. 119 for 1974 and UN (1983) p. 47 for 1980.

RAM in the first half of the 1970s. Later in the 1970s, however, the Japanese industry produced a large share of 4k and 16k dynamic RAM; in the early 1980s it became the largest producer of 64K dynamic RAM, and of 256K dynamic RAM, as Table 6.8 shows.

Table 6.7 Production of integrated circuits as share of total production of semiconductors

	1974	1978	1982
	%	%	%
Europe	25	30	53
United States	52	75	83
Japan	33	48	67

Sources: For 1974: Pottier and Touati (1981), p. 56 and interviews. For 1978 and 1982: UN (1983), p. 34.

Table 6.8 Japanese share of the world memory market

	%	
1K RAM	0	(early 1970)
4K RAM	12	(mid-1970s)
16K RAM	40	(1979)
64K RAM	70	(1981)
64K RAM	54	(1984)
256K RAM	90	(1984)

Source: Pollack, *The New York Times*, 2/28/82 and 9/6/84; *The Economist*, 3/15/82; Dataquest in Borrus, Millstein and Zysman (1982), p. 87.

During most of the LSI period, the United States had a positive trade balance in semiconductors (Table 6.9) but a negative trade balance in integrated circuits. shows. This negative trade balance did not represent a loss of competitiveness, however. Rather, it reflected the growth of imports of integrated circuits from American-owned offshore production and assembly facilities.[20] Japan, on the other hand, has had a positive trade balance in semiconductors since 1978 and in integrated circuits since 1980 (see Table 6.9).

Most European countries had a negative trade balance in both transistors and integrated circuits,[21] as Table 6.10 shows. The major exceptions were France (positive trade balance in transistors during the 1970s), and Italy (positive trade balance in analog integrated circuits during the late 1970s).

The evolution of the European semiconductor industry during the LSI period can be divided into two periods: the first half of the 1970s and the second half of the 1970s and early 1980s. During the first half of the 1970s the European industry did not produce standard LSI semiconductor devices on a large scale. During the late 1970s and early 1980s the European industry became increasingly committed to large-scale production of LSI devices.

Table 6.9 United States and Japanese trade balances in semiconductors and integrated circuits

A. United States trade balance in semiconductors, 1970–1981 ($ millions)

	All semiconductors
1970	+260
1971	+191
1972	+140
1973	+229
1974	+286
1975	+250
1976	+293
1977	+151
1978	−152
1979	−201
1980	−223
1981	−40

B. Japanese trade balance in semiconductors and integrated circuits, 1976–1981 ($ millions)

	All semiconductors	Integrated circuits
1976	−84	−142
1977	−4	−91
1978	+85	−53
1979	+119	−4
1980	+366	+253
1981	+349	+239

Sources: A. US Department of Commerce (1979); Electronic Industry Association, Electronics Market Data Book, various years; United Nations (1983). B. MITI and Ministry of Finance (Japan in United Nations, 1983, p. 69).

2.2 *The lagging competitiveness of European producers during the first half of the 1970s*

After the 1970–71 semiconductor recession and the end of the logic war, European semiconductor producers did not commit themselves to the large-scale production of standard LSI devices. They continued to specialize in discrete devices, linear integrated circuits and custom digital integrated circuits.

Philips, for example, continued to produce discrete devices and linear integrated circuits. Because of the world semiconductor crisis, Philips experienced a peak in operational losses in integrated circuits in 1971.[22] Then, in 1972

Table 6.10 Trade balance in transistors and integrated circuits in major European countries, 1970–1982 ($ millions)

	West Germany TR	West Germany IC	France TR	France IC	Great Britain TR	Great Britain IC	Italy TR	Italy IC
1970	−31	na	−2	na	na	na	0	na
1971	−16	na	+24	na	na	na	+7	na
1972	−20	−24	+16	−8	na	na	+5	−1
1973	−26	−77	+22	−19	na	na	+0	+1
1974	−38	−99	+15	−37	−30	−54	−6	−5
1975	−39	−46	+12	−33	−15	na	−5	−7

	West Germany TR	West Germany ICL	West Germany ICD	France TR	France ICL	France ICD	Great Britain TR	Great Britain ICL	Great Britain ICD	Italy TR	Italy ICL	Italy ICD
1976	−54	−41	−58	+13	−14	−28	−20	−67	na	−6	−1	−7
1977	−62	−72	−82	+20	−18	−30	−25	−11	+9	−5	+13	−11
1978	−74	−86	−77	+28	−26	−46	−19	−28	+20	−10	+3	−19
1979	−88	−104	−98	+37	−37	−48	−26	−25	−12	−12	+4	−17
1980	−96	−102	−210	+33	−83	−86	−29	−13	−70	−15	+4	−31
1981	−60	−81	−125	+29	−23	−100	−34	−24	−83	+9	+14	−51
1982	−59	−85	−163	+9	−40	−98	−30	−25	−148	−6	+9	−59

Note: TR = Transistors; IC = Integrated Circuits; ICL = Linear Integrated Circuits; ICD = Digital Integrated Circuits.
Source: Nimexe (various years)

Philips reorganized its semiconductor production by separating its discrete and integrated circuits group and by increasing the specialization of its subsidiaries.[23] During the first half of the 1970s the Nijmegen (Netherlands) plant produced linear integrated circuits and, somewhat later, began a small production of CMOS integrated circuits; Valvo (West Germany) mainly produced linear integrated circuits; RTC (France) produced mainly digital bipolar integrated circuits; and Mullard (Britain) produced linear integrated circuits and began manufacturing MOS memories. In addition, in 1971, Philips acquired a share in the American firm Electronic Arrays, which specialized in MOS integrated circuits.

Following the semiconductor crisis of 1970–71, Siemens, like Philips, continued to produce mainly discrete semiconductors and various types of integrated circuits, such as bipolar linear integrated circuits for the consumer and industrial markets, ECL integrated circuits for the computer market, Schottky TTL for remote control and MOS integrated circuits for control in TV sets.[24] However, the production of such digital integrated circuits remained limited in size. Siemens' production of integrated circuits was part of its electronics components division.[25] During the first half of the 1970s the ratio of the (external) sales of the component division over Siemens' total sales of electronics and electrical products ranged between 4 per cent (1971) and 7 per cent (1974).[26] Siemens' internal sales of electronics components covered an average of 30 per cent of the total sales of electronic components during the first half of the 1970s. In 1975, during the electronic components crisis, internal sales absorbed that proportion of the production of electronic components that could not be sold on the external market. Approximately one-third of the production of electronics components went to consumer applications; the remaining two-thirds went to computer, industrial and telecommunications uses. Between one-fourth and one-third of (external) sales were exported (Siemens-Berichte, various years). During the first half of the 1970s, Siemens also aimed at improving its competitiveness in several electronics final products, such as computers and process automation, and in nuclear power.

Like both Siemens and Philips, AEG-Telefunken did not move into the large-scale production of LSI devices. Even though AEG-Telefunken produced MOS integrated circuits for the consumer and telecommunications markets and, for a short time, 3F microprocessors for its subsidiary Olympia, it produced these on a limited scale:[27] AEG-Telefunken produced mainly discrete semiconductor devices and custom and linear integrated circuits. Moreover, during the first half of the 1970s, between 40 per cent and 50 per cent of AEG-Telefunken's semiconductor production was used in-house mainly for consumer electronics applications.[28]

None of the French firms switched production towards mass production of LSI devices either. During the first half of the 1970s, Sescosem, the major French semiconductor producer, continued to produce mainly discrete devices

8

Table 6.11 Firm shares of the world semiconductor market (percentage of the total market)

	1972	1974	1976	1980	1983
	%	%	%	%	%
Texas Instruments	11.8	12.1	11.4	11.2	8.9
Philips (with Signetics from 1975)	5.1	5.2	6.7	6.6	4.9
Motorola	9.0	9.0	8.0	7.8	8.9
NEC	–	2.2	6.0	5.4	7.7
National Semiconductor	2.2	3.8	4.6	5.5	4.9
Hitachi	3.9	3.6	4.2	4.7	5.9
Toshiba	4.2	3.2	4.0	4.5	5.1
Fairchild	4.8	6.0	5.3	4.0	na
Siemens	2.3	2.6	na	3.0	1.7
Intel	1.3	2.1	2.6	4.1	4.3
RCA	2.3	na	3.2	2.3	na
Signetics	1.4	2.2	(2.2)	(2.7)	(2.0)
Matsushita	2.0	2.8	4.4	2.1	na
ITT	2.8	3.0	2.5	1.7	na
Mostek	na	na	na	na	na
AMD	na	na	na	na	2.8
Thomson-CSF	1.4	1.0	na	1.0	0.7
AEG-Telefunken	–	0.7	na	1.0	0.7
SGS-ATES	2.2	1.5	na	1.2	1.3
Total $ millions	3,450	5,373	5,762	14,119	17,410

Note: na = not available

Sources: For the years 1972, 1974 and 1976: US Department of Commerce (1979). For the year 1980: Dataquest in UN (1983), p. 130. For the year 1983: Integrated Circuit Engineering in *Elektronik*, 10 February 1984; Dataquest in *Elettronica*, May 1984; *Industry Week*, 2 July 1983 and interviews.

and linear integrated circuits. However, Sescosem was unprofitable and, in 1972, was absorbed by its parent company Thomson-CSF (Pottier and Touati, 1981, p. 55). During this period, Sescosem's importance declined from 23 per cent of the French semiconductor market in 1968 to 16.3 per cent in 1974.[29] The other major producer in France, Radiotechnique Compelec (RTC) (Philips's subsidiary), continued to produce mainly discrete devices and bipolar integrated circuits.

As a consequence of the specialization of French domestic producers in discrete devices, France had a very low ratio of integrated circuit production/ semiconductor production in 1974. In the French industry, in fact, integrated circuits accounted for 23 per cent of total semiconductor production, vs. 33 per cent for Japan and 52 per cent for the United States.

Table 6.12 European semiconductor market

A. Firm shares—1974, 1978 and 1983 (percentage of total market)

	1974	1978	1983
Philips-Signetics	18.0	17.0	13.5
Motorola	8.6	7.0	8.2
Texas Instruments	14.1	12.6	9.7
National Semiconductor	2.6	2.2	4.6
Fairchild	4.0	2.8	na
Siemens	10.8	10.4	7.4
SGS-ATES	5.7	3.9	3.5
NEC	na	na	na
Thomson-CSF	4.4	3.7	3.5
AEG-Telefunken	5.2	4.8	na
ITT	5.5	7.0	3.8
Intel	na	2.3	4.5
RCA	3.6	2.3	na

B. Country shares—1977, 1982 and 1983 (percentage of total market)

	1977 %	1982 %	1983 %
American firms	50	53	50
European firms	48	40	40
Japanese firms	2	7	10
Total	100	100	100

Note: na = not available

Sources: A. For 1974: SGS-ATES; for 1978: SGS-ATES in *Mondo Economico* (12/20/80); for 1983: Dataquest in *Elettronica*, May 1984. Interviews. B. Dataquest in *Elettronica*, October 1983 and *Electronic Times*, 1/5/1984.

During the first half of the 1970s, British semiconductor producers also stayed away from the large-scale production of standard digital integrated circuits. Rather, they occupied market 'niches' by producing custom integrated circuits.[30] GEC, for example, decided to exit from the production of semiconductors in 1969 and, during the first half of the 1970s, was only marginally involved in semiconductor production (Sciberras, 1980). Plessey, on the other hand, concentrated its semiconductor production on a few types of products for a limited number of markets: it succeeded in this effort and in 1973 it developed the very high speed bipolar Process III integrated circuit (Sciberras, 1977). Ferranti chose to specialize in the production of specific TTL integrated circuits rather than to enter the production of memories. Ferranti was also successful and developed the collector diffusion isolation (CDI) process for bipolar devices. Finally, Lucas produced specific types of semiconductors, mainly for in-house use in vehicles and aircraft equipment: in 1972 it developed

Table 6.13 European integrated circuit market

A. Firm shares—1980, 1981 and 1983 (percentage of total market)

	1980	1981	1983
Philips-Signetics	10.9	14.1	11.5
Motorola	8.6	8.0	7.5
Texas Instruments	12.9	11.7	11.8
National Semiconductor	7.3	7.1	6.3
Fairchild	4.7	4.9	na
Siemens	8.6	9.1	5.2
SGS-ATES	5.1	7.0	3.8
Plessey	1.9	2.4	na
NEC	na	na	4.4
Thomson-CSF	1.7	1.7	3.7
AEG-Telefunken	2.0	2.4	na
ITT	3.6	3.7	na
Intel	7.3	6.7	6.3

B. Country shares—1975 and 1977 (percentage of total market)

	1975 %	1977 %
American firms	78	64
European firms	22	33
Japanese firms		3
Total	100	100

Note: na = not available
Sources: A. Dataquest in *Business Week*, 28 June 1982 and in *Elettronica*, May 1984. Interviews. B. Dataquest in *Elettronica*, October 1983.

an MOS-ROM for fuel injection systems. In fact, Lucas used 60 per cent of its semiconductor production in-house.

As a consequence of these developments, these British firms had less than 15 per cent of the British semiconductor market by the mid-1970s (Sciberras, 1977). Rather, this market was dominated by American firms such as Texas Instruments, Fairchild and Motorola, and by ITT and Philips (through its subsidiary Mullard).

In Italy STET (a state-owned telephone and electronics holding of IRI) purchased SGS in 1971 and merged it with ATES, STET's semiconductor producer. At the time of the merger, SGS produced some LSI devices for the computer market. The production of LSI devices, however, represented only a small part of SGS's total production. SGS was mainly producing discrete semiconductors (approximately 50 per cent of its total sales) and integrated circuits for the computer and telecommunication markets.[31] ATES, on the other hand, was mainly producing germanium power devices for the consumer market.

Although it was smaller in size,[32] ATES had more influence than SGS over decisions concerning future productive specialization, in part because the world semiconductor crisis had aggravated the already unprofitable situation at SGS. Therefore SGS-ATES decided to become increasingly specialized in products for the consumer market, and to concentrate its attention on power devices and linear integrated circuits.[33] As a consequence of this choice, SGS-ATES introduced power linear integrated circuits for radio and TV receivers and for automotive products, becoming a European leader in this product range. SGS-ATES also continued to produce discrete semiconductors, although it eliminated its R&D in this field.

The integration of SGS-ATES with STET was a formal (not effective) integration.[34] SGS-ATES's sales were directed towards the external market and, only to a limited extent, towards STET's in-house market. STET, in fact, was not committed to supporting a large-scale, in-house production of semiconductors, in part because its in-house demand for semiconductors was limited. Therefore, the integration of SGS-ATES into STET remained mainly a 'financial' integration.[35]

2.3 *The restructuring and commitment of major European firms to the production of LSI devices*

Since the second half of the 1970s, the European industry had become increasingly committed to the large-scale production of LSI devices for several reasons. First, after the 1974–5 semiconductor crisis, European producers became convinced that they needed to produce LSI devices in order to retain their overall shares in the European semiconductor market. Discrete and linear integrated circuits were becoming less and less important and LSI integrated circuits were becoming more and more important in both the European and the overall world semiconductor market. Second, European producers realized the significance of the increasing digitalization of linear functions in semiconductor devices and the increasing interdependence among the computer, telecommunication and consumer markets. These trends required that these European producers establish capabilities in LSI integrated circuits. Third, European governments increasingly supported their domestic semiconductor industries, particularly in their efforts to produce LSI semiconductor devices. These European governments thought that a domestic productive capability in LSI devices was of strategic value for reasons of national security and for sustaining the electronic and the manufacturing industries as a whole. The commitment of European producers to LSI production was completed by the early 1980s; by that time all of the most important European producers (Philips, Siemens, SGS-ATES, Thomson-Brandt, INMOS) has begun to produce LSI semiconductor devices.

Philips concentrated its R&D and production activities in LSI devices on microprocessors, custom and semicustom devices and MOS and CMOS integrated

Table 6.14 Firm shares of the German, French, British and Italian semiconductor markets (percentage of total market)

	Germany			France			Great Britain			Italy		
	1974[a]	1978[b]	1983[e]	1974[a]	1978[b]	1983[e]	1973[c]	1977[c]	1983[e]	1974[d]	1978[b]	1983[e]
Philips	16	15		17	14		17	18		5	10	
Siemens	22	21		2	na		na	na		3	5	
AEG-Telefunken	11	9		na	na		na	na		na	na	
Thomson-CSF	2	na		14	16		na	na		4	5	
Intel	na	na		na	4		na	8		na	na	
GEC	na	na		na	na		na	6		na	na	
Lucas	na	na		na	na		na	5		na	na	
Plessey	na	na		na	na		4	3		na	na	
Ferranti	na	na		na	na		5	1		na	na	
SGS-ATES	3	3		na	6		2	3		16	15	
ITT	7	7		3	na		14	8		4	9	
Texas Instruments	12	13		14	12		18	22		10	14	
Motorola	7	6		11	12		14	11		7	10	
Fairchild	2	na		5	3		na	na		13	13	
National Semiconductor	1	na		3	4		na	5		na	na	
European firms			42			41			33			39
American firms			48			52			53			53
Japanese firms			10			7			14			8

Note: na = not available

Sources: (a) SGS-ATES; (b) Dosi (1981); (c) Sciberras (1978); (d) Fast (1980); (e) Dataquest in *Elettronica*, November 1984.

164   *The large-scale integrated period*

Table 6.15 US patents granted to major European firms, 1969–June 1981
(percentage over total patents granted)

	Semiconductors	Materials	Junctions	Housing	Transistors	Integrated circuits	Digital integrated circuits	Central processing unit
Philips	4.3	3.0	5.6	2.4	2.5	6.1	3.0	1.0
Siemens	3.1	2.2	3.0	3.7	5.4	2.3	2.5	0.7
Thomson-CSF	0.6	–	0.9	–	0.8	0.3	0.5	–
AEG-Telefunken	0.5	–	0.5	0.3	0.5	0.9	–	–

Source: OTAF.

circuits after 1978. Each of its European subsidiaries specialized in some specific products. Philips's plants in Nijmegen were producing CMOS and bipolar (mainly linear) integrated circuits for consumer applications; and Valvo (Philips's German subsidiary) was producing mainly linear integrated circuits for the German consumer electronics industry. During the late 1970s/early 1980s, however, Valvo began producing LSI devices (mainly NMOS integrated circuits) and several lines of Intel's microcomputers under licence.[36] Radiotechnique Compelec (RTC) (Philips's French subsidiary) was producing bipolar integrated circuits (TTL, ECL, I2L) while Mullard (Philips' British subsidiary) was producing linear integrated circuits, NMOS integrated circuits for teletext and viewdata, ROM memories, and microprocessors for television applications. Finally, Philips' Swiss subsidary was producing CMOS integrated circuits for consumer applications.

In 1975, in order to obtain technological capabilities in LSI technology, Philips purchased Signetics, the ninth largest American semiconductor producers. At the time of purchase, Signetics produced bipolar integrated circuits —mainly TTL—and was starting production of MOS integrated circuits. Following the purchase, however, Philips directed Signetics to continue to produce bipolar integrated circuits and to put more effort into MOS integrated circuits.[37]

Philips's commitment to LSI device production was not profitable at the beginning. Because Philips committed itself to the production of LSI semiconductor devices much later than American producers, its production of integrated circuits remained unprofitable except in 1979.[38]

A relevant share of the LSI devices produced by Philips was used internally, together with Philips' production of discrete semiconductors and linear integrated circuits. During the late 1970s Philips absorbed an average of 50 per cent of its semiconductor production for in-house uses and satisfied between 80 and 90 per cent of its semiconductor needs from in-house sources. The internal consumer market was much larger than the industrial and telecommunication markets: in fact, by 1980, television and radio receivers, and other audio-visual products, constituted more than 26 per cent of Philips's total sales. Consumer products required linear and digital integrated circuits: digital integrated circuits were used for remote control, digital tuning, and the increasing digitalization of the television receivers.[39]

Table 6.16 US Patents granted to Philips, 1969–1981 (percentage over total patents granted)

	Semiconductors	Materials	Junctions	Housing	Transistors	Integrated circuits	Digital integrated circuits	Central processing unit
1969	3	2	6	–	1	2	–	–
1970	6	2	7	5	7	7	2	–
1971	6	3	6	4	6	8	2	–
1972	8	4	7	6	8	9	5	–
1973	6	7	9	1	3	8	2	7
1974	4	7	4	5	–	4	4	–
1975	8	6	9	3	10	9	5	–
1976	7	–	10	9	2	10	3	2
1977	3	–	4	–	–	5	6	1
1978	4	4	4	–	–	7	6	–
1979	6	8	5	3	7	6	8	–
1980	3	6	3	2	–	na	na	na
1981	8	14	12	–	–	na	na	na

Note: na = not available
Source: OTAF.

During the 1970s and early 1980s, Philips also continued to maintain an innovative record in discrete devices and integrated circuits for the consumer market. In 1970, for example, it developed an analog integrated circuit for TV receivers; this development was spurred by the demand of Philips's video division for such an integrated circuit.[40] Philips also developed colour decoder integrated circuits for TV receivers and integrated circuits for teletext applications and Prestel systems.[41] As Tables 6.15 and 6.16 show, between 1969 and 1981, Philips had 4.3 per cent of the total American patents granted in semiconductors, and was particularly specialized in semiconductor junctions and linear integrated circuits.[42]

During the second half of the 1970s, Siemens, like Philips, gradually committed to LSI device production, because Siemens' top management understood the crucial importance of these devices for Siemens' competitiveness in electronics final products, especially for telecommunication and industrial equipment.[43] Siemens' integrated circuit division had been experiencing losses since 1965.[44] Following the semiconductor crisis of 1975, Siemens decided to commit resources to the production of digital integrated circuits.[45]

Siemens opened plants and research centres in Germany and Austria. It concentrated on MOS integrated circuits, microprocessors, memory devices and integrated circuits for the external consumer market and the internal telecommunication and computer markets.[46] It absorbed approximately 50 per cent of its semiconductor needs and between 80 and 90 per cent of its microprocessor needs through internal production.

In order to acquire technological expertise, Siemens developed a follower-imitative strategy.[48] In 1977 it purchased 17 per cent (later 20 per cent) of the stock of the American semiconductor producer Advanced Micro Devices (AMD).[49] Through this purchase, Siemens acquired a share in a firm that specialized in semiconductors for the telecommunication, industrial and computer markets, that produced reliable products and that second-sourced Intel's and Zilog's microprocessors.[50] In addition, Siemens purchased Microwave Semiconductor,[51] 80 per cent of Litronix[52] and 60 per cent of the stock of Advanced Micro Computer (AMC), a joint venture between Siemens and AMD for the development of microcomputers.[53] Siemens also made technological agreements with Intel and obtained licences for the production of the Intel 8080 microprocessor in 1976; in the following years it produced several types of Intel microprocessor.[54] Finally, Siemens obtained a licence from the American firm Rockwell to produce magnetic bubble memories[55] and co-operated with Fujitsu in the production of electronic components[56] and with Philips in the production of integrated circuits for the consumer and power device markets.[57]

The follower-imitative strategy proved successful in the early years, but unsuccessful when a new world semiconductor crisis occurred. In fact, while in the years 1977–9, Siemens' sales of integrated circuits to the external market were profitable, after 1980 they became unprofitable.[58]

In the early 1980s, therefore, Siemens modified its strategy: it continued to commit resources to standard LSI products, but it also decided to put more emphasis on custom and equipment-specific integrated circuits. By concentrating on application-orientated chips, Siemens intended to exploit the advantages of vertical integration and of close interaction between semiconductor producers and users. This advantage was closely linked to the 'leading customer' concept in the development and production of integrated circuits.[59] A large share of these custom and equipment specific integrated circuits was used in-house by Siemens.

The renewed commitment of Siemens to standard LSI semiconductor devices production was related to the need to acquire a general technological capability in LSI. Siemens concentrated on both microprocessors and memories. Siemens manufactured 16k RAMs and Intel microprocessors.[61] It also invested at first

Table 6.17 US patents granted to Siemens, 1969–1981 (percentage over total patents granted)

	Semiconductors	Materials	Junctions	Housing	Transistors	Integrated circuits	Digital integrated circuits	Central processing unit
1969	5	6	9	6	1	3	2	–
1970	4	–	2	7	9	1	1	–
1971	2	5	2	2	2	–	1	–
1972	3	–	6	6	5	2	1	–
1973	5	3	3	4	24	2	7	–
1974	3	–	–	7	10	2	1	–
1975	3	–	–	9	8	3	3	–
1976	2	–	1	7	7	3	4	1
1977	4	–	4	5	7	4	6	2
1978	7	16	4	3	17	5	5	–
1979	3	4	4	–	2	4	6	1
1980	5	3	6	5	2	na	na	na
1981	2	–	5	–	–	na	na	na

Note: na = not available
Source: OTAF.

DM150 m. ($83m.) and, later on, DM100m. ($55m.) in new production facilities for the production of 16k and 64k RAMs.[62] In 1980 its sales of integrated circuits, including microprocessors and memories (totalling approximately $250m.) constituted 40 per cent of its semiconductor sales.[63]

During the late 1970s and early 1980s, Siemens continued to be competitive in a wide range of other types of semiconductor devices, which were either technologically more mature or had a less dynamic growth than LSI devices. Siemens remained highly competitive in opto-electronics devices (mainly light-emitting diodes) and in power semiconductor devices. Thus, in 1980, for example, Siemens introduced a power MOS semiconductor device, SIPMOS, that was initially used for 200V and later for 500V.[64]

During the second half of the 1970s, AEG-Telefunken continued to be a major producer of discrete semiconductor and opto-electronic devices, and entered the production of custom LSI-integrated circuits.[65] In 1980, AEG-Telefunken's production of integrated circuits amounted to DM100m. ($55m.),[66] approximately one-fifth of its production of electronics components. Digital devices accounted for only 20 per cent of the production of integrated circuits: they consisted mainly of custom MOS integrated circuits.[67] In 1980 AEG-Telefunken began producing microprocessors with a licence from Mostek.

The choice of types of LSI device to be produced was determined to a great extent by AEG-Telefunken's internal market. AEG-Telefunken absorbed approximately 30 per cent of its semiconductor production and satisfied less than 10 per cent of its semiconductor needs with internal production.[68] This internal market came mostly from consumer electronic applications; it absorbed discrete devices and linear integrated circuits. Like both Philips and Siemens, AEG-Telefunken was able to innovate incrementally in semiconductor devices for final electronics products in the area—consumer goods—in which the firm specialized, for example: AEG-Telefunken introduced a linear bipolar integrated circuit radio, a remote control and colour processing integrated circuit, and a linear tuner for TV sets.[69]

In general, AEG-Telefunken's semiconductor production was negatively affected by the firm's crisis which exploded in 1974 and which continued throughout the 1970s and early1980s. The crisis of AEG-Telefunken originated in the early 1970s from its nuclear power operations. These operations absorbed a large amount of AEG-Telefunken's capital and were unprofitable. AEG-Telefunken eventually sold these operations to Siemens in 1976. In addition, AEG-Telefunken's success in various consumer products declined in the mid-1970s. These two situations drained internal funds that could eventually have been used in semiconductor activities.

As a result of this crisis, in late 1982 and early 1983 AEG-Telefunken's semiconductor activities were restructured. In late 1982 Telefunken Electronics was established. Telefunken Electronics was a joint venture between AEG-Telefunken (49 per cent), United Technologies (49 per cent) and the Süddeutsche Ind-Beteiligungs Gmbh (2 per cent). In early 1983, Telefunken Electronics made another joint venture with United Technologies: Eurosil Electronics Gmbh. Eurosil Electronics Gmbh, a former German producer of custom integrated circuits, was a joint venture between Telefunken Electronics (43.6 per cent), United Technologies (43.4 per cent) and the Diehl Gruppe (13 per cent). These two joint ventures were focused on the production of custom NMOS and CMOS integrated circuits and of gate arrays.

In France Thomson began large-scale production of LSI devices in the late 1970s. After the 1975 semiconductor crisis, with sales on the French market falling by 20 per cent, Thomson-CSF established second-source agreements with Motorola to produce microprocessors and with Fairchild to produce integrated circuits.[70] In 1976 Thomson-CSF purchased 35 per cent (later increased to 65 per cent) of EFCIS. EFCIS was a small firm that had been founded in 1972 by CEA[71] in order to begin the production of custom integrated circuits for the industrial and military markets. After Thomson-CSF took it over, however, EFCIS became the major French producer of MOS integrated circuits and received substantial support from the French Integrated Circuits Plan. EFCIS produced both NMOS and CMOS integrated circuits, 8-bit 6800 and 16-bit 68000 Motorola microprocessors, EPROM-EEPROM memories, and custom integrated circuits. In 1980, in fact, custom integrated circuits accounted for 37 per cent of EFCIS's production; microprocessors and memories for 33 per cent; systems for 13 per cent; and integrated circuits for telecommunication and office products for 17 per cent.[72] EFCIS's production for the internal market decreased from 50 per cent in 1978 to 40 per cent in 1980,[73] however; this reduction resulted from the decreasing weight of custom integrated circuits in EFCIS' production.[74]

Thomson-CSF's internal market formed by telecommunication, industrial and military equipment, and by the consumer products of Thomson-Brandt, absorbed a large share of Thomson's LSI devices. In general, Thomson-CSF absorbed approximately 25 per cent of its semiconductor production and satisfied approximately 50 per cent of its requirement for discrete devices and 10 per cent of its requirement for integrated circuits from internal sources.[75]

In the early 1980s, Thomson-CSF became the major French producer of LSI devices and, despite significant losses, renewed its commitment to LSI devices production. In early 1983 Thomson-CSF purchased Eurotechnique, a joint venture between Saint-Gobain and National Semiconductor created in 1978 with the support of the French Government. Therefore, in 1983, Thomson-CSF had the control of Sescosem, EFCIS and Eurotechnique and a production ranging from discrete semiconductors to integrated circuits. In this year, with the support of the Government, Thomson-CSF decided to commit resources to LSI device production: microprocessors, memories of a certain type, custom and semicustom integrated circuits. According to Thomson-CSF intentions, these LSI devices did not have to be mainly directed to in-house consumption, but had to be sold on the external competitive market.

Two other French firms vertically integrated into LSI production in the late 1970s: Matra and Schlumberger. Matra established a joint venture with the American producer Harris in 1978 and by the early 1980s became the second largest French producer of LSI devices. It produced mainly CMOS, NMOS and bipolar integrated circuits. Schlumberger (a company which provided services for oil extraction) acquired the American producer Fairchild. Fairchild had been unsuccessful in many different product lines (especially consumer products) and was not at the technological frontier in MOS integrated circuits. After its acquisition by Schlumberger, however, Fairchild focused its R&D and productive efforts on MOS integrated circuits and RAM memories.[76]

The reasons why these French producers of final products entered into semiconductor production are closely related to the evolution of technology. Matra and Schlumberger decided to develop an in-house production of semiconductors in order to reap the benefits of technological and productive integration; Matra produced defence equipment and, therefore, needed an increasing number of sophisticated semiconductor components; Schlumberger used computerized tools for measurement and oil research and, therefore, required increasing amounts of semiconductor components.[77]

In Britain GEC tried to re-enter the market for standard digital integrated circuits in 1978; it established a joint venture with Fairchild for producing memories and microprocessors. This joint venture ended, however, in 1980, after Fairchild was bought by Schlumberger. GEC then decided not to enter the large-scale production of standard digital integrated circuits. Rather, by joining separate subsidiaries in 1981, it established Marconi Electronic Devices Limited (MEDL), which mainly produced power devices and hybrid circuits and, to a lesser extent, custom bipolar and CMOS integrated circuits. GEC used part of its production of semiconductor devices in its military and telecommunication products and sold the rest on the external markets.[78]

Plessey, on the other hand, remained a producer of custom semiconductor devices but increased its commitment to LSI devices after 1980. Before 1980 Plessey's production of semiconductor devices was mainly directed to the consumer and the military markets. Since 1980, because of the increasing strategic importance of custom LSI devices for its final products, Plessey has concentrated its efforts on specialist and custom LSI devices for the telecommunication and the military markets. Plessey produced CMOS and high-speed bipolar integrated circuits, and semi-custom integrated circuits. An increasing share of Plessey's semiconductor devices was absorbed in-house.[79]

Since the mid-1970s Ferranti decided to select a few technological 'niches' and to focus its effort on these. Because of its pioneering activity in uncommitted logic arrays for semi-custom devices, and its technological capability in bipolar technology, Ferranti specialized in LSI devices for market niches not entered by American producers. In particular, Ferranti concentrated on bipolar semicustom integrated circuits.[80]

Ferranti 'niche' strategy in LSI devices proved successful. The firm's sales of integrated circuits grew from £1.6m. (1976) to £35m. (1983) and firm profits increased. In the early 1980s, Ferranti had a 25 per cent share of the world uncommitted logic arrays semicustom market.[81]

Most of Ferranti's production was sold on the external market. Only approximately 6 per cent of this production went to the internal military, industrial and system equipment production.[82] The rest of Ferranti's production of bipolar semi-custom devices was absorbed by the external computer and telecommunication markets.

While Ferranti and Plessey have been concentrating on the custom and semicustom markets, a new British firm, INMOS, has been focusing on the standard LSI market. INMOS was established in 1978 by the British government through the British Technology Group (the former National Enterprise Board). INMOS opened two plants, one in Britain and one in the United States, and began production in 1981–2. It concentrated its efforts on the memory and microprocessor markets.[83] During the early years INMOS was unprofitable and had to be supported by government funds. By early 1984 the government had provided £65m. in cash and £35m. in loan guarantees, most of which were used for the plant in the United States.[84] In 1984, however, INMOS became profitable and was purchased by the British electronics firm Thorn.

In 1982 a major change in the structure of the British semiconductor industry occurred; ITT sold the majority of STC and of its British semiconductor operations, to the British. Thus, since 1982, a new British vertically integrated semiconductor firm (STC) has been operating in the British industry. This firm produces memory devices and integrated circuits for the telecommunication and the military markets.[85]

In Italy SGS-ATES entered the large-scale production of LSI devices in the late 1970s/early 1980s. Its entry followed the difficulties faced by the company as a result of the world semiconductor crisis and related also to the financial difficulties of the STET group in the mid-1970s. These latter difficulties stemmed from the losses of SIP (STET's telephone common carrier) which had drained finances away from SGS-ATES at the time (1974–6) of great need and which had interrupted the internal transfer of funds from SIP to the electronics producers of the STET group through inter-company loans and pricing.[86]

SGS-ATES's LSI production was begun as a result of the company's prediction of high growth rates of markets for these devices and of the policies of STET and Olivetti. The entry into LSI production coincided with a reduction in the range of products produced by SGS-ATES. STET supported SGS-ATES's restructuring plan and promoted management changes for this purpose. STET also fostered R&D collaboration and purchases between SGS-ATES and other STET firms (Selenia and, in particular, Sit-Siemens—later Italtel).[87] While the R&D collaboration of SGS-ATES and Sit-Siemens did materialize, the amount of SGS-ATES's semiconductor devices purchased by Sit-Siemens remained limited, both in absolute and in relative terms. Sit-Siemens, however, induced SGS-ATES to focus on the production of LSI devices for telecommunication equipment. In addition, Olivetti, in its change from electromechanical to electronics office equipment production,[88] needed increasing amounts of LSI semiconductor devices. During the second half of the 1970s, in fact, Olivetti purchased between 5 and 10 per cent of SGS-ATES's sales. During this period Olivetti also collaborated with SGS-ATES in some R&D projects.[89]

SGS-ATES concentrated its efforts in specific types of integrated circuits: standard MOS-LSI devices (particularly CMOS and NMOS integrated circuits for the telecommunication, automotive, and consumer markets), ROM, EPROM and EEPROM, and microprocessors[90] for the computer, office equipment, telecommunications equipment and consumer markets.[91] In addition, SGS-ATES continued to produce power transistors for the industrial market and linear integrated circuits for the consumer and industrial markets. SGS-ATES remained highly competitive in these product areas.[92]

Like other European producers, SGS-ATES followed a policy of technological co-operation and of second source with major semiconductor producers. In 1977 it established second-source agreements with Zilog and Fairchild for the production of the Z80 and Z8000 microprocessors and the F8 microprocessor, and in 1982 it established a technological agreement with the Japanese producer Toshiba.

As a result of the change in policy and of the increase in world semiconductor demand, in 1983 SGS-ATES became profitable after years of losses. The return to profitability was associated with a rapid growth of the firm.

During the late 1970s and early 1980s, two Swedish vertically integrated firms, Rifa and HAFO, began producing custom LSI devices. Rifa was part of Ericsson, the Swedish telecommunication producer; it produced custom LSI devices for the Ericsson Information Systems division, established in 1982 for the production of office and telecommunication equipment. HAFO was part of the ASEA group, the Swedish industrial equipment producer; it produced custom LSI devices for both in-house use and external customers.

Finally, during the 1970s, other European semiconductor producers continued to produce either discrete semiconductors, power and opto-electronic devices, or integrated circuits for specific markets and applications. Brown Boveri, Semikron, and Ansaldo fall into the first category: Brown Boveri and Semikron produced mainly discrete devices and Ansaldo produced mainly power devices. Bosch, Eurosil and MEM-Ebauches, on the other hand, fall into the second category. Bosch (West Germany) produced integrated circuits for automobiles, while Eurosil (West Germany) and MEM-Ebauches (Switzerland) produced integrated circuits for watches and other consumer applications.[93] In addition, the semiconductor divisions of other electronics firms produced a wide range of products. For example, Piher in Spain produced discrete devices and bipolar integrated circuits and Favag in Switzerland (owned by Hasler) produced bipolar integrated circuits.

2.4 *American direct foreign investments, European acquisitions and international co-operation*

During the LSI period, the interaction between the European, the American and the Japanese industries occurred not only through American (and, later on, Japanese) direct foreign investments in Europe, but also through European acquisitions of American merchant producers and through the R&D co-operation of European firms with American and Japanese firms.[94] While direct foreign investments in Europe continued a trend established in previous periods, acquisitions and co-operation were new phenomena. They were the result of the commitment of European firms to LSI technology and to their inability to master fully and immediately such a complex technology.

The direct foreign investments of American firms in Europe, begun during the integrated circuit period, continued during the LSI period. For example, in the first half of the 1970s, Motorola and National Semiconductor opened plants and design centres in Scotland. Later in the 1970s, Japanese direct foreign investments were made in Europe: NEC and Fujitsu opened plants and assembly operations in Ireland.

These direct foreign investments were the result of technological, protectionist and locational factors. American firms during the whole LSI period, and Japanese firms in the late 1970s and early 1980s, maintained technological and commercial advantages in LSI technology over European producers. These firms fully exploited these advantages by locating production facilities in Europe, and therefore profited both from the 'proximity' advantage previously discussed,[95] and from the circumvention of the tariff barriers protecting European markets.

The acquisition of American merchant producers by European firms and the co-operation among firms of different countries took place during the second half of the 1970s and the early 1980s. As previously mentioned, these acquisitions and participations were the result of the decision of cash-rich European electronics final goods producers to integrate into LSI devices—considered crucial for their competitiveness at the electronics final goods and equipment level.[96] Either because these firms did not want to run the risk of, and could not afford the time necessary for, the establishment of an in-house capability, or because they wanted to supplement their in-house capability, they took control of highly competitive American merchant producers. During the second half of the 1970s, for example, Philips took control of Signetics (1975); Siemens took control of Microwave Semiconductor (1979) and purchased a 20 per cent participation in Advanced Micro Devices (1977); Ferranti took control of Interdesign (1977); Lucas purchased a participation in Siliconix (1977); Schlumberger took control of Fairchild (1979) (see Table 6.18).[97]

Co-operation among European, American and Japanese firms in basic research or in the development of specific products took place in the early 1980s.[98] For example, Siemens collaborated with Philips in basic R&D in semiconductors, and with Intel for the development of specific types of microprocessors; Philips collaborated with Signetics and Motorola for the development of specific types of microprocessors; SGS-ATES and Toshiba collaborated for the development of specific types of MOS integrated circuits.

These co-operations were the result of the increasing complexity and appropriability of LSI technology, of the rising cost of doing R&D, together with the increasing technological capability of European firms in LSI technology.[99] By co-operating with other firms, European firms shared the cost, reduced the risk and avoided the duplication of R&D.[100]

3 Demand factors: the convergence of electronics markets

During the 1970s, the size and structure of the European demand for semiconductor devices underwent major changes. The overall size of this demand

Table 6.18 Corporate investments in American semiconductor companies by
European corporations, 1975–1982

Semiconductor producer	Acquiring company	Percentage owned
Advanced Micro Devices	Siemens	20
American Microsystem	Bosch	25
Fairchild	Schlumberger	100
Unitrode	Schlumberger	15
Interdesign	Ferranti	100
Litronix	Siemens	80
Signetics	Philips	100
Siliconix	Lucas	24
Solid State Scientific	VDO Schilling	25
Microwave Semiconductor	Siemens	100
Databit	Siemens	100
Threshold Technology	Siemens	100
Semiprocesses	CIT-Alcatel	25

Note: na = not available
Sources: Hazewindus (1982); Borrus, Millstein and Zysman (1982), p. 34; Braun and
MacDonald (1982), p. 176; UN (1983), p. 265.

increased because of the increase in the demand for electronics final goods
and because of the increase in the semiconductor content of the various
electronics final goods, as Table 6.19 shows. The structure of this demand
changed because of the convergence of the computer, telecommunication
and electronics consumer goods markets and because of the increasing importance of the telecommunication and computer demand relative to consumer
demand (see Tables 6.20–6.23).

As a result of the convergence of electronics final markets and the increasing
digitalization of linear functions, a high European demand for LSI devices
materialized in the second half of the 1970s and the early 1980s. This demand
came not only from the traditional computer producers, but also from small
computer producers, telecommunication equipment producers, and electronics
consumer goods producers. This demand substituted part of the demand for discrete devices and linear integrated circuits of European electronics final goods
producers. These changes in the structure and type of demand and their effects
on the European semiconductor industry are examined in the following pages.

During the LSI period, the European computer demand for semiconductors
increased at a very high rate. Computer sales had a high rate of growth in Europe
during the 1970s because of their pervasiveness and because of the establishment
of a new type of computer: the small computer. Mini, small business, micro and
personal computers had a very high rate of growth during the second half of
the 1970s and the early 1980s.

Table 6.19 Semiconductor intensity of final products (value of semiconductors
as percentage of the value of the final products)

	1970	1973/74	1975	1980
Computers	1.5		4	6
Main-frame		3–4		
Mini-computers		15–20		
Consumer-Auto			4	6
TV—Colour		8–10		
Pocket calculators		50–70		
Telecommunications		2–3		
Industrial			1	2
Government/Military			2	3

Source: Gnostic Concept in Fast (1980) and SGS-ATES (1974).

American computer producers continued to dominate the European computer market. In 1983, for example, they had 81 per cent of the European computer market.[101] IBM maintained the largest market share in Europe. DEC became a leader in minicomputers. Hewlett Packard and Honeywell controlled large shares of the European computer market (see Tables 6.24, 6.25 and 6.26). In the late 1970s, these firms were joined by minicomputer producers such as Commodore, Apple and Tandy.[102]

During the 1970s European producers continued to be not very successful in the European computer market.[103] This was the case with Siemens, in general purpose computers: until the early 1970s, Siemens produced computers with an RCA licence; but RCA soon exited from computer production (in 1971) (OECD, 1977). In 1971–2 Siemens joined CII and Philips to form the Unidata Company. This union lasted only a few years (1971–5), however, and was not successful (OECD, 1977). During the second half of the 1970s Siemens co-operated with Fujitsu, producing computers that were plug-compatible with IBM systems.[104] AEG-Telefunken also had an unsuccessful performance in the computer market. In the early 1970s AEG-Telefunken was joined by Nixdorf in Telefunken Computer Gmbh, which produced large computers for technical and scientific operations. This joint venture proved unsuccessful, however: in 1974 Computer Gmbh was purchased by Siemens.[105] Following this failure, AEG-Telefunken continued to produce only small computers for industrial control.[106] In France CII and Honeywell-Bull[107] merged (in 1975) and the mini computer division of CII merged with a computer subsidiary of Thomson to become La Société Européene de Mini-Informatique et de Systèmes (SEMS). These new companies did not prove successful, however (OECD, 1977). In Britain, the major domestic producer, ICL, held a relatively large share of the domestic computer market in the early 1970s. ICL followed a strategy of independently developing main-frame computers and of developing a wide range

Table 6.20 Size of the electronics final markets: the United States, Europe and
Japan (percentage of total sales)

| | United States | | | |
	1972	1976	1980	1982
	%	%	%	%
Consumer-Auto	16	20	19	20
Communications	7	7	6	6
Computers	33	34	38	40
Industrial	7	10	11	10
Federal	37	29	26	24
Total	100	100	100	100

| | Europe | | | |
	1972	1976	1980	1982
	%	%	%	%
Consumer-Auto	32	36	30	32
Communications	17	20	20	20
Computers	34	28	38	37
Industrial	17	16	12	11
Total	100	100	100	100

| | Japan | | | |
	1974	1976	1980	1982
	%	%	%	%
Consumer-Auto	44	41	34	37
Communications	13	13	13	8
Computers	29	31	38	37
Industrial	14	15	15	18
Total	100	100	100	100

Source: *Electronics* (various years)

of products. ICL was able to survive alongside IBM in main-frame computers.
It performed poorly, however, in small computers and its share of the computer
market continued to decline throughout the 1970s (Grant and Shaw, 1979).
In Italy, not until the mid-1970s, did Olivetti begin to produce small computers,[108] as part of its conversion from mechanical to electronics technologies.[109]

As they had done in the 1960s, European governments tried to support
their domestic computer producers during the 1970s, but they were unsuccessful.

Table 6.21 World end-use markets for semiconductors—1974, 1979 and 1984 (percentage of total market)

	1974	1979	1984 (forecast)
	%	%	%
Consumer-Auto	35	31	30
Computers	20	21	22
Industrial	15	16	16
Distribution	15	15	15
Telecommunications	8	10	11
Other	7	7	6
Total	100	100	100

Source: Motorola in *Electronics*, 30 May 1981, p. 4 and in UN (1983), p. 28.

Their policies were directed towards established domestic producers (ICL, CII-Honeywell-Bull, Siemens and AEG-Telefunken). European government policies used both public procurement and specific programmes. The extent of public procurement varied from country to country. In West Germany, for example, the government controlled between 7 and 13 per cent of the domestic market in 1974;[110] this was accomplished mainly through the purchase of large systems from Siemens and AEG-Telefunken.[111] In Britain the government controlled approximately 12 per cent of the market in 1974, mainly through the purchase of large computers from ICL.[112] In France this figure was approximately 8 per cent in 1974 (US Department of Commerce, 1976). Also, the various computer programmes varied widely among countries. In West Germany the second (1970-5) and third (1976-9) data processing programmes, for a total of $1,325m. and $866m., respectively, supported computer applications and data processing education and training. These programmes put less and less emphasis on hardware,[113] and more and more emphasis on software and applications. Siemens received DM 400m. ($173m.) from the second and DM 280m. ($122m.) from the third computer programme.[114] In France the two Plans Calculs 1966–71 and 1971-6, respectively totalling 726.1m. frs. ($140m.) and 1.315m. frs. ($263m.), mainly went to support the hardware divisions of the computer producers.[115] The British government supported the introduction of a new series of ICL computers between 1971 and 1975 by providing ICL with £40m. In addition, the government supported the domestic industry through the Advanced Computer Technology Project, the Software Product Scheme, the Application Programmes and the Microelectronic Support Scheme (OECD, 1977, p. 247). Finally, in Italy the government prepared a sectoral plan which was included in the more general 'Programma Finalizzato Elettronica' (Electronics Plan) (Ministero dell'Industria, 1979). Most of these programmes did not reach their goals and European producers continued to lag behind American producers.

Table 6.22 End-use markets for semiconductors: the United States, Europe and Japan (percentage of the total market)

	United States		
	1973	1975	1980
	%	%	%
Consumer-Auto	25	23	28
Computers	28	33	44
Telecommunications	} 29	29	10
Industrial			10
Government	18	15	8
Total	100	100	100

	Europe				
	1973	1976	1978	1982	1984
	%	%	%	%	%
Consumer-Auto	32	31	30	23	27
Computers	22	19	20	16	20
Telecommunications	12	14	14	15	23
Industrial	16	17	18	16	22
Military/Space	5	6	5	5	8
Distribution	13	13	13	25	na
Total	100	100	100	100	100

	Japan
	1980
	%
Consumer-Auto	52
Computers	32
Telecommunications	10
Industrial	6
Total	100

Source: For the United States: US Department of Commerce (1979) for 1973 and 1975; ICE (1981) for 1980. For Europe: SGS; Fast (1980) for 1973, 1976 and 1978; *Electronics* (1983) for 1982; Dataquest in *Elettronica*, November 1984 for 1984. For Japan: Japan Electronics Almanac in UN (1983).

Because of their unsuccessful performance and because of the lack of competitive LSI European producers of standard LSI devices during the first half of the 1970s, European computer producers faced a dilemma similar to that faced during the 1960s. They could either buy 'state-of-the-art' LSI devices from American firms, or purchase less competitive LSI devices from European

Table 6.23 End-use markets for integrated circuits: world, the United States, Europe and Japan (percentage of total market)

	World		
	1970	1976	1980
	%	%	%
Consumer	5	15	28
Telecommunications	5	10	} 54
Computers	64	45	
Industrial	} 26	30	10
Government			8
Total	100	100	100

	United States	
	1974	1978
	%	%
Computers	40	35
Consumer-Auto	18	12
Telecommunications	5	8
Industrial	9	11
Government	14	9
Distributors	14	25
Total	100	100

	Europe	
	1978	1980
	%	%
Consumer-Auto	45	28
Computers	18	23
Communications	15	18
Industrial	22	26
Government	na	5
Total	100	100

	Japan	
	1974	1980
	%	%
Consumer-Auto	58	46
Computers	25	38
Communications	} 17	10
Industrial		6
Total	100	100

Source: World: Truel (1980) in Ernst (1981), p. 76 and UN (1983), p. 28. United States: US International Trade Commission (1979). Europe: 1978, A. D. Little (1980) in Truel (1981), p. 9. 1980, UN (1983), p. 31. Japan: 1974, Japan Electronic Industry Development Association (1980) in Pugel-Kimara-Hawkins (1983). 1980, UN (1983), p. 30.

Table 6.24 Major computer producers in the world, 1981 (sales of computers and information systems) ($ millions)

American	
IBM	26,340
DEC	3,587
European	
ICL	1,513
Olivetti	1,436
CII-Honeywell-Bull	1,353
Siemens	1,330
Nixdorf	856
Japanese	
Fujitsu	1,950
NEC	1,330
Hitachi	1,290

Source: Datamation and Computopia in UN (1983).

Table 6.25 Firms shares in the minicomputer and the personal computer markets —1974 and 1984

	Minicomputer Europe 1974	Personal Computer World 1984
	%	%
DEC	19.2	3.6
Hewlett-Packard	6.4	2.9
IBM	5.6	20.6
Honeywell	5.1	na
Apple	—	18.4
NEC	—	6.8
Olivetti	—	1.8
Others	63.7	45.9
American producers	51.2	na
Other producers	48.8	na

Note: na = not available

Sources: For Europe: Sciberras *et al.* (1978), p. 22. For World: Dataquest in *Markt-Technik*, 15 June 1984.

Table 6.26 American, European and Asiatic computer markets

A. Firm shares of the American and the European computer markets—
1973–1975 (percentage of the total market)

	United States 1975	Europe 1973	Europe 1975	France 1975	Great Britain 1975	West Germany 1975	Italy 1975
	%	%	%	%	%	%	%
IBM	69	53	54	55	40	62	69
Honeywell	9	na	na	15	10	na	9
CII	na	na	na	10	na	na	na
Siemens	na	3	4	na	na	17	na
ICL	na	11	9	na	31	na	na
Others	22	33	33	20	19	21	22
Total	100	100	100	100	100	100	100

B. Country shares in the American, the European and the Asiatic markets
—1983

	USA	Europe	Asia
	%	%	%
American companies	97	81	45
Japanese companies	2	2	53
European companies	1	17	2

Note: na = not available

Sources: A. For United States, France, Great Britain and West Germany: Rosner (1978). For Italy: OECD (1977). For Europe: *Business Week*, 17 December 1979. B. *Business Week*, 16 July 1984.

firms, or produce less competitive LSI devices in-house. In order to survive in the computer market, most European producers chose the first option.

As a result, just as in the 1960s, no large-scale computer demand for European LSI devices materialized during the early 1970s. Computer producers purchased 'state-of-the-art' LSI devices from American firms or from European subsidiaries of American multinational corporations. Therefore, the high growth rate of the European computer market during most of the 1970s did not transmit a significant pull on the European production of LSI devices.

During the second half of the 1970s, the high growth rates of the small computer market and the convergence of the electronics final markets affected the structure and the conduct of the European computer industry. Former specialized European computer producers modified their strategies. For example, in the early 1980s ICL put more emphasis on alliances with other firms to develop new products[116] and CII-Honeywell-Bull took control of La Société Européenne de Mini-Informatique et de Systèmes (SEMS). Other companies, such as Olivetti, Nixdorf and Sinclair, entered the booming small computer and electronics office equipment market. Finally, because of the convergence of electronics final markets, diversified electronic final goods producers, such as Philips and Siemens, decided to commit resources to the production of specific types of computers and office equipment in order to remain competitive in their traditional electronics consumer goods and telecommunication equipment markets. Through various programmes of support, several European governments helped the commitment of established electronics producers and of new firms to the new type of computer market.

The changing structure of the European computer industry during the late 1970s and the commitment of electronics firms to computer and office equipment production resulted in a growing demand for standard and custom LSI semiconductor devices, such as memories, microprocessors and logic circuits. LSI devices continued to be at the heart of mainframe computers, but they also became crucially important in the new types of electronics final goods.

As a result of this increase in demand and in the importance of LSI devices, in the late 1970s and the early 1980s European producers decided to establish an in-house capability in LSI devices, while continuing to buy 'state-of-the-art' LSI devices from American semiconductor firms. This was the case with respect to Philips, Siemens and Thomson-CSF. As previously discussed, in some cases (for example, Philips and Siemens) the beginning of an in-house production of LSI devices was supplemented by the acquisition of (or participation in) a highly competitive American merchant producer.

Yet the increasing demand for, and the strategic importance of the LSI devices did not result in the entry of new European merchant producers into the LSI device market because of the lack of potential demand for this group of producers. This lack of potential demand was the result of the choice of European computer producers to buy 'state-of-the-art' LSI devices from American merchant producers rather than to purchase non-competitive LSI devices from would-be new European merchant producers. These new producers therefore did not have the time required for the establishment and accumulation of technological capability, R&D and productive experience necessary to reach the technological frontier in LSI devices. On the other hand, as previously seen, for strategic and long-term considerations, some European electronic final goods producers had the commitment and financial resources to subsidize the establishment of an in-house capability for LSI devices.[117]

The convergence of electronics final markets affected the European telecommunication market differently. In this market, European telecommunication equipment producers were internationally competitive, supported by their governments and vertically integrated into discrete devices and linear integrated circuits production. The convergence of the telecommunication equipment market with the computer and electronic goods markets, and the digitalization of linear functions, greatly increased the requirements of LSI devices, which were going to substitute discrete devices and linear integrated circuits. Increasing digitalization meant that digital semiconductor components replaced linear components in switching, that optical fibres for pulse code modulation were used more in transmission, and that microprocessors and LSI devices became important components in terminal equipment (see OECD, 1981). As a result, the global semiconductor content of telecommunication equipment increased.[118]

Table 6.27 Major telecommunications equipment producers in the world (percentage of total market shares, 1982)

	%
Western Electric	23
ITT	9
Siemens	9
Ericsson	5
GTE	5
Northern Telecom	5
NEC	4
GEC	4
Thomson-CSF	3

Source: Arthur D. Little in *Business Week*, 24 October 1983.

In addition, telecommunication demand for semiconductors increased because of the high level of investments made by telecommunication service providers during the 1970s and the early 1980s. During the 1970s in France, for example, the General Direction of Telecommunications (DGT) invested heavily in new telecommunication equipment in order to modernize the French telecommunication system; and in Italy STET (the electronics public holding) made large investments during the first half of the 1970s (Antonelli and Lamborghini, 1978). In the early 1980s, France, West Germany, Britain and Italy continued to invest in new electronic types of telecommunication equipment.[119]

A large part of this demand for telecommunication equipment continued to be directed to domestic producers. In 1976, for example, common carriers accounted for 73 per cent of the demand for telecommunication equipment in West Germany, 72 per cent in France, 77 per cent in Britain and 86 per cent in Italy (OECD, 1981, p. 114). In the same year, domestic producers (non-foreign owned) accounted for 75 per cent of the West German market, 78 per cent of the French market, 76 per cent of the British market and 71 per cent of the Italian market (OECD, 1981).[120]

During the 1970s, however, in several European countries a more flexible system of purchasing, a less conservative attitude among common carriers and a decrease in the degree of protection of domestic producers developed. In

Britain the PTT (Post, Telephone and Telegraph)[121] moved away from the Bulk Supply Arrangement (a system of market-sharing agreements) toward more competitive tenders (OECD, 1981). In France, the DGT changed its procurement practices for more competitive biddings in order to avoid becoming dependent on ITT for equipment and in order to favour purchases from CIT-Alcatel (CGE subsidiary) and Thomson-CSF (Ergas, 1983). In France the DGT switched from a conservative attitude in the 1960s to a more innovative one in the 1970s. This change of attitude brought with it the development of the E-10, a digital electronics switching system. The E-10 was produced by CIT-Alcatel with the support of CNET (National Centre for Studies on Telecommunications), part of the DGT (Diodati, 1980; Ergas, 1983). It was introduced in 1970 and produced in 1974. In Britain the 'Telecommunication Bill' of November 1980 created British Telecom, which pursued de-regulation of the whole sector. Since then, British firms have been subject to an increasingly competitive pressure. The decrease in the degree of protection is also related to the increasing share of new, unprotected types of demand, as in the case of the PABX (Private Branch Exchange).[122]

In order to maintain their competitiveness in this new and less protected market, European telecommunication producers had to use state-of-the-art LSI devices rather than absorb their established in-house production of discrete devices and linear integrated circuits. Like the European computer producers, European telecommunication producers could not afford to buy non-competitive LSI devices from would-be European merchant producers or to use discrete devices and linear integrated circuits. In the first case, their telecommunication equipment would not perform as well as that of their international competitors; in the second case, their transition to fully digital telecommunication equipment would have been slowed down. Therefore, while at the beginning European telecommunication equipment producers purchased state-of-the-art LSI devices from American firms (as did the European computer producers), with time, several of these producers integrated vertically in the production of LSI devices.

The convergence of electronics final market and the increasing digitalization of linear function also affected the electronic consumer goods market. In this market, European firms continued to be competitive. In the colour TV market, for example, Philips, Thomson-Brandt, Grundig, and AEG-Telefunken, were market leaders.[123] European producers were successful because they were protected by a 14 per cent tariff, they manufactured good quality picture tubes in-house, and they produced TV receivers for transmission standards (the German Pal and the French Secam) that were incompatible with American and Japanese standards (Levy, 1981). However, in the late 1970s and early 1980s, several of these producers faced increasing competition on the European consumer market from Japanese producers.[124] This new competition was, in part, a result of the convergence of different electronics final markets and, in part, a result of the increasing digitalization of electronics consumer products.

While, during the first part of the 1970s, the consumer market required mostly discrete devices and linear integrated circuits, during the second half of the 1970s and the early 1980s it required increasing amounts of LSI devices. These LSI devices were mostly of the MOS and CMOS type.

As in the telecommunication market, these changes in the consumer demand for semiconductors played an important role in the beginning of LSI production in Europe. Several consumer producers, in fact, such as Philips, AEG-Telefunken and Thomson-Brandt, which had to use large amounts of LSI devices in their consumer goods, began producing these devices in-house.[125]

4 Supply factors: entry through vertical integration

During the LSI period, the European semiconductor industry continued to be composed of the same group of vertically integrated receiving tube producers which had been dominating it since its beginning: Philips, Siemens, Thomson-CSF, AEG-Telefunken. At the beginning of the 1970s, these firms mainly specialized in discrete devices and linear integrated circuits; by the early 1980s, most of them were committed to the production of custom or standard LSI devices. The rest of the industry was composed of vertically integrated non-receiving tube producers (such as SGS-ATES, Ferranti, Plessey, ITT) and of a few merchant producers (INMOS and EFCIS, recently purchased respectively by Thorn and Thomson-CSF).

The change from a specialization in discrete devices and linear integrated circuits in the early 1970s to a commitment to the production of LSI devices in the early 1980s was the result of the convergence of the final electronics markets and of the increasing digitalization of linear functions, examined in previous sections. This commitment to LSI technology was the result of the technological beliefs of European electronics firms about the increasing importance of LSI technology and occurred through the upstream integration of these producers into standard or custom LSI devices production. These technological beliefs and this upstream integration are examined in the following pages.

By the second half of the 1970s, most European electronics final goods firms agreed both that LSI devices were strategically important for innovation at the electronics final goods level and that LSI devices would be used increasingly in electronics final goods.[126] Their beliefs were based on observations accumulated over the preceding years relating to the effects of the application of microprocessors, memory devices and LSI logic circuits in electronics final goods. These beliefs were confirmed and reinforced by a series of reports, studies and publications: by the late 1970s, in fact, the future trends in LSI devices and in their applications were understood and predicted with a certain accuracy.[127]

Following these technological beliefs, several European electronics final goods producers decided in the second half of the 1970s and in the early 1980s to integrate into the production of LSI semiconductor devices. The incentives for European electronics final goods producers to integrate upstream into the R&D and production of LSI devices, however, were very different in the case of custom and standard LSI.

In the case of custom LSI, electronics final goods producers were provided with two types of incentives: the integration of components and systems and the technological interrelationship between electronic final products. By integrating upstream into the production of custom LSI devices, electronic final goods producers avoided the high transaction costs involved in the exchange of custom inputs when human and environmental factors were present; profited from better communication between producers and users within the corporation and were able to reap economies of scope in the R&D and production, particularly when the final goods producer was a diversified firm.[128] It should be noted, however, that many of the producers who vertically integrated into the production of custom LSI devices produced only limited quantities of these devices.

In the case of standard LSI devices, on the other hand, electronic final goods producers were provided with a very different set of incentives to integrate upstream: the need for a continuous supply of standard LSI devices, coupled with the fear that the overall growth of the supply of these devices would not be able to keep up with the overall growth of demand. A continuous supply of LSI devices was required for the co-ordination of the supply of LSI devices with the production of electronics final goods. The more the LSI devices became complex and entire sub-systems, the less they became interchangeable with other types of LSI devices. As a consequence, shortages of specific types of LSI devices affected the production of electronic systems during the LSI period more substantially than shortages of standard integrated circuits had affected the production of electronic systems during the integrated circuit period. It is also important to note that the fears that shortages would occur or that demand would outstrip supply were associated, in the LSI period, with a large, developed industry that could not keep up with demand. They were very different from the ones that had appeared in the transistor period, which were associated with the ability of electronics final goods producers to deal with an infant industry that had not developed sufficiently to keep up with demand. Although these fears about the possibility of demand outstripping supply were common to both American and European corporations, additional pressure was present in the European case as a result of the possibility of cut-offs from American supply during periods of peak demand. European producers were heavily dependent on the supply of American state-of-the-art devices and knew that American producers preferred to supply American customers before European customers in cases of excess demand. During the 1970s, in fact, shortages of certain state-of-the-art LSI devices during periods of peak demand occurred more frequently in Europe than in the United States.

For late entrants such as the European corporations, however, the beginning of LSI devices production proved difficult to accomplish successfully. Since LSI technology was highly complex and appropriable[129] and technological advance in LSI increasingly cumulative, early commitment and relevant experience became key factors of success in LSI technology. While European electronics final goods producers had the financial means and the human resources for the beginning of the LSI production, they had neither a developed and advanced technological, design and software capability in LSI semiconductors nor the cumulative manufacturing experience of the American and Japanese LSI devices producers.

Given these disadvantages in terms of time and experience compared to American and Japanese firms, the European firms committed to establishing an in-house LSI devices capability followed an articulated strategy. This strategy included the beginning of an in-house production of LSI devices cross-subsidized internally from other profitable activities of the corporations; the acquisition of American merchant producers at the frontier of LSI technology; the co-operation with American firms in well-defined R&D projects and areas; and the search for support from national governments.[130] It should be noted that these co-operation policies mainly involved the leading LSI devices producers. Co-operation with leading LSI devices producers guaranteed European firms a high technological input, which could eventually offset the risk of leaks of proprietary information.[131]

In the innovation of custom LSI devices, European LSI producers could exploit some specific advantages from their vertically integrated structure and from their competitiveness in telecommunication equipment and consumer electronics. Because they were vertically integrated, these producers were able to place more emphasis on design, on complexity, on uniqueness, and on system characteristics; their main goal remained end-product reliability and performance. This approach differed from that of merchant producers who aimed at producing LSI devices at minimum cost. In addition, they were able to focus their semiconductor R&D on the specific needs of their internal customers, use the idiosyncratic language of the internal customer and acquire applied knowledge and feedback on the quality of their semiconductor devices by the internal customer; as a result, user uncertainty was greatly reduced. Innovativeness in custom semiconductor devices, however, required that both the final goods producer and the semiconductor producer within vertically integrated firms were competitive producers. Only in this case, in fact, could the advantage of having an internal customer be translated into increased innovativeness at the LSI device level.

5 Public policy factors: the support of the European industry

During the 1970s, and particularly during the late 1970s, European governments began to support their domestic semiconductor industries through large and sector-specific policy measures. The German government since 1974, the French and British governments since 1978, and the Italian government later on, have supported R&D and subsidized the investments of private firms in semiconductors.

European governments' support for their domestic semiconductor industries represented a major change from the lack of support of the 1960s. Compared to those years, in fact, recent European government support was larger, more focused and more concentrated on the semiconductor industry. It was only in the late 1970s, therefore, that public policy became an important factor in the European industry.

In most European countries government policies were initiated as a result of the recognition of the strategic and economic importance of the semiconductor industry and as a reaction to the realization that the domestic semiconductor industries were technologically inferior to their American counterpart. Because of the strategic and economic importance of this industry, the technological dependence on the United States and the lack of international competitiveness in semiconductors seriously impaired the technological capability and the economic performance of these countries.

European government policies were therefore aimed at reducing the existing technological lags of domestic producers and at developing a productive capability in the most advanced technological areas. These policies focused on LSI semiconductor technology, as the technology at the heart of developments in the semiconductor, the computer, the telecommunication, and the electronics industrial equipment industries.

The policies, however, varied across countries and were applied to domestic industries which differed in terms of their actual competitiveness, the size of their markets and the degree to which they had been penetrated by foreign producers. German policy reacted quickly to perceived needs, allotted a large amount of money to domestic firms, used a broad range of policy tools and was consistent and clear; it consisted mainly of R&D support and investment subsidies. By contrast, French, British and Italian policies were established later than German policies (second half of the 1970s), and were not as consistent and clear. The policies of these three countries consisted of both R&D and production refinement support and involved the direct intervention of the government in the structure of the domestic industries. The French and British policies, however, were larger than the Italian policy in terms of funds effectively given to (not only planned for) domestic semiconductor firms.

5.1 *West Germany*

German government policy towards the semiconductor industry relied on the market mechanism and set general guidelines regarding the area of intervention. The reliance on the market mechanism was rooted in the German post-war tradition of public policy, and so was the relative freedom given to firms to pursue their technological objectives in the way they considered best. However, the implementation of government programmes and the results obtained were monitored by independent experts. At the completion of the projects, the results were made public and the products were made commercially available.

As a consequence, government R&D support did not push firms in technological directions which they would not have taken independently as a result of their technological beliefs. Rather, government R&D support permitted firms either to undertake risky and expensive projects or to continue research in areas which had proved unsuccessful in the past and which would eventually have been abandoned by these firms.[132] Particularly before 1978, government R&D support induced a clustering of projects in some types of research where government money was available.[133]

The ministries of Research and Technology (BMFT) and of Economics were the main government agencies in charge of decisions about guidelines and areas of intervention. The PTT, on the other hand, maintained a conservative attitude towards the introduction of new components and equipment, while the military exerted a marginal influence on the industry.

German government policy gave funds to a limited number of firms: Siemens, AEG-Telefunken, Valvo (since 1977) and Intermetall (since the early 1980s). Support was mainly given to domestic producers, and only in most recent times to subsidiaries of multinational corporations with R&D and production facilities in West Germany, such as Valvo (Philips) and Intermetall (ITT). During the 1970s, for example, Siemens was supported for approximately one-fifth of its R&D expenditures in integrated circuits.[134]

German government policy towards the semiconductor industry during the LSI period may be separated into three periods: early 1970s–1974, 1974–8, and 1978–84. The first period is characterized by limited support for the domestic semiconductor industry. The second period is characterized by the first major European programme in support of the semiconductor industry. The third period is characterized by a series of other programmes of support.

The first period (1970–74) was characterized by limited but increasing support for the R&D and production of semiconductors in domestic firms within the framework of other existing programmes: the Second Electronic Data Processing Programme (1971–5) and the programmes on New Technologies and Electronics run by the German Ministry of Research and Technology (BMFT). Such support increased after 1972 as a direct consequence of the European defeat in the semiconductor crisis and the price war of 1970–71:

German firms did not perform satisfactorily in integrated circuit production and lagged behind American producers. As a result, they exerted pressure on the government to give them support. The government was convinced that a strong domestic semiconductor industry was important both for reasons of national security and for the technological progressiveness of the whole of German industry. The German government provided the industry with DM 9.8m. ($2.8m.) in 1971, DM 31.7m. ($9.9m.) in 1972 and DM 40.5m. ($15m.) in 1973 for R&D in electronics components (Federal Republic of Germany, BMFT, 1974). Within the electronic components sector, the German government specifically supported R&D in integrated circuits— DM 0.6m. ($171,000) in 1971, DM 8.2m. ($2.5m.) in 1972 and DM 12.0m. ($4m.) in 1973—and in opto-electronics— DM 0.5m. ($143,000m.) in 1971, DM 6.1m. ($1.9m.) in 1972 and DM 7.3m. ($2.7m.) in 1973.

There were also several programmes which supported R&D in semiconductors. From 1971–3 the Data Processing Programmes allotted DM28.3m. ($8m.) for R&D in circuits and DM 16.5m. ($4.7m.) for R&D in fast memories; the space programme granted DM 1.0m. ($285,000) to R&D in semiconductors and DM 0.2m. ($57,000) to R&D in integrated circuits, while the Ministry of Defence granted DM27.55m. ($7.8m.) to R&D in solid-state physics, electronic components and opto-electronics devices.[135] In addition, the DFG (German Research Society) gave DM4.9m. ($1.4m.) for research in semiconductor electronics, DM 0.6m. ($170,000) for research in opto-electronics and DM 33.3m. ($9.5m.) for basic research (BMFT, 1974).

During the 1970–74 period, a share of government support was directed towards discrete devices, in which German firms were already competitive internationally. Support for power devices increased from DM1.9m. ($543,000) (1971) to DM3.7m. ($1.2m.) (1972) to DM4.3m. ($1.6m.) (1973); support for thick and thin film circuits and hybrid circuits increased from DM0.3m. ($93,000) (1972) to DM2.5m. ($926,000) (1973) to DM4.5m. ($1.7m.) (1974) while support for transistors and diodes went from DM0.5m. ($143,000) (1971) to DM3.1m. ($968,000) (1972) to DM2.5m. ($926,000) (1973). Government support for power devices, thick and thin film circuits, hybrid circuits, transistors and diodes, contributed to the slowness with which German firms switched from the production of discrete and power devices to the production of digital integrated circuits. However, the 1970–74 government support was based on programmes prepared in 1969, a year in which the crucial importance of digital integrated circuits was not yet fully understood by the industry and a clear awareness of how technological developments would have proceeded was absent.

This understanding and this awareness were present in the policies of the second period (1974–8), which were characterized by the adoption of a specific programme for electronics components by the BMFT (BMFT, 1974). In late 1971 this programme had been preceded by a set of co-ordinated R&D programs in semiconductors submitted by Siemens and AEG-Telefunken to the Ministry of Research and Technology. At that time, both the Ministry of Research and Technology and the German semiconductor producers became convinced that it was necessary to develop a specific programme for electronic components and, in 1974, the Electronics Component Programme (1974-8) was launched. The goals of this programme were to establish an efficient domestic components industry, to initiate co-operation in R&D between universities, research institutions and industry in a few selected priority areas, and to concentrate R&D in particular research projects that were, as of yet, unimportant for domestic producers from an economic point of view.[136] A total of DM376m. ($106m.) was given to German firms from 1974 to 1978 under this programme (Fusfeld and Tooker, 1980, p. 50).

The Electronics Components Programme concentrated on five priority areas: integrated circuits throughout the whole range of MOS, linear and digital, bipolar and custom integrated circuits—DM85.5m. ($32.9m.) in 1974-8; opto-electronics throughout the whole range of light-emitting devices and light-receiving devices for camera tube techniques—DM 62.0m. ($23.8m.); materials, in particular high-purity monocrystalline silicon, III-V compounds and materials with magnetic, dielectric and ferro-electric properties—DM 36.0m. ($13.8m.); automated manufacturing processes—DM28.5m. ($10.9m.); basic research, particularly in new semiconductor technologies[137] and new measurement techniques—DM56.0m. ($21.5m.).[138] The Programme discontinued support, however, for semiconductor power devices, thick and thin film circuits, hybrid circuits, transistors and diodes. It justified this elimination by the fact that German firms were already competitive in these products.

Under the Electronics Components Programme, government funds normally covered 50 per cent of the cost of industrial R&D projects or, in exceptional cases in which risk was very high, 80 per cent of the cost of the project. Government funds allotted for basic research in non-profitable institutions covered 100 per cent of the basic research project; most of these funds were given to DFG, the German Research Society.

Between 1974 and 1977 the German government also channelled DM 39.7m. ($15.2m.) into the R&D in circuits and fast memories through the Second Data Processing Programme; DM 0.25m. ($96,000) into the R&D in semiconductors through the programme on Space Research Technology; and DM54.46m. ($20.9m.) into the R&D on the reliability and performance of semiconductors through the Federal Ministry of Defence (Fusfeld and Tooker, 1980).

Whereas up to 1978 government programmes concentrated more on basic research and applied research than on the development and introduction of new products in the market, the situation changed around 1978 as a result of the unsuccessful performance of German producers in the integrated circuit market. After 1978 a large number of government projects financed the introduction and application of new products into the market. Some of these programmes were directed toward small and medium-size firms as well. R&D support and investment subsidies to the semiconductor industry are given in Table 6.28. German programmes set two main goals regarding microelectronics: the improvement of the competitiveness of domestic producers in VLSI technology and the support of small and medium-size firms in their application of microelectronics.[139] In order to reach the first goal, a special VLSI plan was established; this plan devoted DM 35m. ($19.4m.) in 1979, DM 42m. ($23.3m.) in 1980 and DM 48m. ($20.9m.) in 1981 to support R&D in VLSI technology (Fusfeld and Tooker, 1980). In order to reach the second goal, the government allocated a large share of the funds from these programmes to small and medium-size firms. In 1981, for example, DM 23.4m. ($10.2m.) were given to such firms for R&D support.[140] These programmes were part of more general programmes regarding the electronics industry as a whole. They included the Third Data Processing Programme, the Information and Documentation Programme, the Technical Communication Programme and the Electronics Components Programme. These programs covered basic technology, products, systems and their applications, and the production, transmission and treatment of information.[141] In addition, the VDI-Technologiezentrum in Berlin and other institutions were put in charge of advising small and medium-size firms on the application of microelectronics technology.[142]

Table 6.28 R&D support and investment subsidies in the semiconductor industry in West Germany, 1979–1982

	1979 DMm. $m.	1980 DMm. $m.	1981 DMm. $m.	1982 DMm. $m.
R&D support	90.5 50.2	90.6 50.3	96.0 41.7	103.0 42.4
Investment subsidies	19.1 10.6	15.6 8.6	20.6 8.9	18.0 7.4

Source: BMFT (1981).

In 1984 the BMFT launched a major programme called Informationstechnik, regarding not only semiconductors, but also other types of electronics components, telecommunications, electronics consumer goods, computers, industrial automation and industrial equipment. The four-year programme (1984–8) includes support for components for peripheral equipment—DM320m. (approximately $112m.); for computer-aided design for integrated circuits—DM 90m. (approximately $32m.); for key electronic components—DM90m. (approximately $32m.). In particular, the programme supports R&D in sub-micron technologies for VLSI semiconductor devices—DM600m. (approximately $210m.); in semiconductor technology—DM200m. (approximately $70m.) and in opto-electronics devices—DM90m. (approximately $32m.).[143]

5.2 *France*

French government policy towards the semiconductor industry was centralized and *dirigiste*. The French government and bureaucracy attempted to direct the technological and productive efforts of the domestic semiconductor producers. This type of policy was possible because of the strength and authority of the French public bureaucracy.[144]

As a consequence, French policy relied heavily on structural measures and on specific directives to domestic producers. French public policy favoured acquisitions among domestic firms and encouraged joint ventures with foreign firms. It also pushed firms to specialize in specific technological areas suggested by the government.

During the 1970s, French public policy towards the semiconductor industry was affected by the rivalry between the DIELI of the Ministry of Industry and the Department of Defence, on one side, and the DAII of the Direction Générale des Télécommunications (DGT) on the other. While the former supported Sescosem (Thomson), the latter was in favour of the establishment of new French producers of integrated circuits. The effects of the DAII-DGT policy position resulted in the creation during the 1970s of new French producers of integrated circuits, first EFCIS, and subsequently Matra and Eurotechnique.[145]

Notwithstanding these tensions, French government policy directed funds mainly to Thomson-CSF. In the early 1970s, funds were channelled to Sescosem and subsequently to EFCIS (controlled by Thomson-CSF) and Sescosem; and finally, after the lack of success of the 1978–82 integrated circuit plan, to Eurotechnique (acquired by Thomson-CSF), EFCIS and Sescosem.

French government policy towards the semiconductor industry can be separated into two periods—the period before 1978 and the period after 1978. The first period was characterized by a series of programmes not specifically directed to the integrated circuit industry. The second period includes several programmes aimed at supporting the domestic semiconductor industry.

During the period before 1978, funds for semiconductor producers were channelled through the VI Plan (1971-5), the VII Plan (1976-81) and the 'Plan Calcul', which was directed towards the domestic computer industry. The VI Plan allocated 1.290m. frs. ($235m.) to the electronic components industry; however, only 700m. frs. ($127m.) were actually used.[146] The VII Plan allocated 430m. frs. ($90m.) to support industrial R&D; 220m. frs. ($46m.) to public R&D in active electronic components and 1.200m. frs. ($250m.) for investment subsidies in electronic components.[147] Most of these funds went to Sescosem (Thomson-CSF), although they were not part of a coherent design for the development of a domestic integrated circuit industry. In fact, only after the French semiconductor industry was affected by the semiconductor crisis of 1974-5 did both the government and the firms become convinced that there was a need for concentrated action in the area of digital integrated circuits.

Hence, in the second half of the 1970s, the government created a plan for integrated circuits (1978–82). The main goal of this plan was to create a French manufacturing capability in LSI and VLSI technology and to make the French industry internationally competitive; this goal was particularly important because domestic producers were almost completely absent from this technology at the time the plan was drafted. The plan was controlled and managed by the DIELI (Direction des Industries Éléctroniques et Informatiques) and was to allocate 350m. frs. ($78m.) over a five-year period to create a laboratory within CNET (National Centre for Studies and Research of the Ministry for Post and Telecommunications) to conduct research on integrated circuits for telecommunications.

The Plan also reorganized the French industry according to well-defined semiconductor technologies and electronic final markets,[148] and provided domestic producers with a 'minimum' of 600m. frs. ($133m.) to support R&D and investment subsidies. It allocated 100m. frs. ($22.2m.) over a five-year period for R&D support of the semiconductor division of Thomson-CSF:[149] this division was to produce bipolar integrated circuits for consumer electronics applications. It also gave 100m. frs. ($22.2m.) over a five-year period to Radiotechnique-Compelec (RTC) (subsidiary of Philips) for R&D support; RTC was required to produce high-speed bipolar integrated circuits, particularly ECL and TTL circuits, for the computer and telecommunication markets and was to become Philips' European subsidiary specialized in these types of products.[150] EFCIS was also assigned 200m. frs. ($44.6m.) over a five-year period for R&D support; EFCIS was required to specialize in MOS technology, and particularly custom NMOS, SOS and CMOS. Eurotechnique was granted 180m. frs. ($50m.) in investment subsidies and low interest loans over a five-year period.[151] Eurotechnique was required to specialize in MOS technology (in particular CMOS and NMOS) for a wide range of applications. Finally, the Plan assigned 120m. frs. ($27m.) in investment subsidies over a five-year period to Matra Harris Semiconductors (MHS);[152] MHS was required to specialize in MOS technology for a wide range of applications.[153] The plan also intervened directly in the structure of the domestic industry. The government fostered the increase of Thomson-CSF's shares in EFCIS,[154] and supported the joint ventures of Eurotechnique and MHS.

Although it was the first major attempt by the French government to support the domestic integrated circuit industry, the French Integrated Circuit Plan has been the subject of much criticism. First, it took one and a half years to complete the plan, and so the plan lagged behind the rapidly changing situation in the world semiconductor industry. Second, the amount of support allocated was limited compared to the amount required by the French industry to reduce its lag behind the American industry through R&D and capital investments. Third, the Plan also kept research centres such as CNET and LIETI (as well as similar centres at major universities) separate from production centres (Lorenzi and

LeBoucher, 1979). Fourth, the Plan chose to make technological agreements with American producers rather than to acquire direct control of some of these American firms, as did Philips and Siemens; these technological agreements exposed French firms to delays and limitations in the flow of technological know-how from American producers.[155] Finally, and more importantly, the plan supported too many producers, so that economies of scale could not easily be obtained by each firm.

In addition to the funds given by the Integrated Circuit Plan, however, domestic semiconductor producers received various funds from government agencies, such as FSAI and DATAR, totalling 400m. frs. ($83m.). In addition, the Plan *Informatisation de la Société* gave these producers 10–15m. frs. ($2.2–3.3m.) on a yearly basis for microprocessor applications.[156]

In 1983 a new integrated circuit plan was launched by the socialist government. This plan was part of the more general plan for the *Filière électronique* (1982–7) in support of the whole French electronics industry.[157] It intended to improve the negative French trade balance in integrated circuits and create a French capability in LSI devices. The support for the domestic semiconductor industry over the period 1982–6 is of 6bn. frs. ($820m.); 3.4bn. frs. ($465m.) are for R&D support; 2.2bn. frs. ($300m.) are for investment subsidies. The Ministry of Industry is the main source of funds. The Department of Defence, the DGT and the Ministry of Research are other sources of funds.[158] The new 1982–6 integrated circuits plan concentrated its support on Thomson-CSF and Matra. Thomson-CSF has become the major French producer, being composed of Sescosem, EFCIS and Eurotechnique. Matra is the second largest French producer. Both firms have been nationalized and, together with CGE and CII-Honeywell-Bull, represent the four major poles of government intervention in the electronics industry.

5.3 *Britain*

British government policy towards the semiconductor industry has some characteristics of the German policy and others of the French policy. If the British tradition, based on the market mechanism, has inspired several British programmes, other programmes have been more 'interventionist'. Considered in overview, British policy during the 1970s and the early 1980s has not been consistent. British policy has been characterized both by support given to firms with no interference in their technological decisions, and at the same time by direct structural intervention in the industry. In the first case, the British government supported the R&D efforts of domestic and foreign subsidiaries. In the second case, it established the public corporation NEB (the National Enterprise Board) in 1975 and the firm INMOS in 1978.

Poor co-ordination existed between the Department of Industry on one side, and the Ministry of Defence, the British Technology Group (BTG) and the

Science and Engineering Council, on the other—all responsible for policy towards the industry.[159] It must be remembered that BTG (formerly NEB) had been established with the task of promoting efficiency and competitiveness in the British industry.

The British government's support for the industry was addressed mainly to domestic producers such as GEC, Ferranti and Plessey. Foreign subsidiaries obtained support in some specific cases, examined below.

British government support for the semiconductor industry during the 1970s can also be divided into two periods: the period before 1978 and the period after 1978. In the first period, the government gave the semiconductor industry limited R&D support and investment subsidies. In the second period, some major programmes were established.

During the period before 1978, the focus of government policy was directed mainly towards the computer industry; semiconductors were, therefore, neglected. Compared to the amount of support given to the domestic industry by the German government, the funds received by the semiconductor industry in Britain for R&D support and investment subsidies were quite limited.

Before 1978 the major government programme had been the Microelectronics Support Scheme: this programme allocated £12m. to support the R&D of domestic producers in microelectronics. The programme was to finance R&D, which would not have been carried out in the absence of such support. The only beneficiaries of this fund were GEC, Ferranti and Plessey. In addition, through the Advanced Computer Technology Project, the government provided an average of £0.8m. per year to domestic semiconductor producers. Then, in 1976, a Component Industry Scheme was established by the Department of Industry in order to support R&D in electronics components. In the period 1976-9 £5m. was given to capital support and approximately £15.5m. was assigned to R&D support in components. This programme was widely spread among twenty-two groups of electronics components. Of the total £20m. it assigned 54 per cent of the total funds to R&D support, 37 per cent to investment subsidy and 9 per cent to restructuring and rationalization support. Semiconductors had the largest share of support: approximately £5m.[160]

During the period 1971-8, the CVD[161] of the Ministry of Defence supported the R&D activities of GEC, STC (ITT), Plessey, Mullard (Philips) and Ferranti. These five firms obtained more than 75 per cent of the total number of contracts assigned by CVD. Of these firms, GEC received more than one-quarter of the total number of research contracts of CVD between 1971 and 1976;[162] its place as major contractor was taken by Plessey between 1975 and 1978, however (Dickson, 1983).

When, in 1974-5, British producers were negatively affected by the world semiconductor crisis, both domestic firms and the government became aware of the *de facto* absence of British producers from the most technologically dynamic parts of the semiconductor industry. The government prepared a series of reports on the state of the industry and discussed a number of public programmes for support.

After 1978 British policy became more comprehensive, both in the scope and the type of intervention; it fostered the application of semiconductors; it supported R&D and subsidized investment of firms and it intervened in the structure of the domestic industry by establishing a new public enterprise (INMOS); it increased the interaction between the Department of Industry and the Ministry of Defence.

However, as Dickson notes, although government policy became more comprehensive during this period, it was often discontinuous; this is particularly true in the case of the military. Moreover, support was stretched over too broad a range of projects and was focused on R&D activities that were not always concerned with commercial goals (Dickson, 1983).

Between 1978 and 1982, the British government launched two major plans in support of the semiconductor industry, and several support programmes. The Microelectronics Support Programme was established first. It granted £70m. (later cut to £55m. by the Conservative Government) for R&D support and investment subsidies to semiconductor producers located in Britain. It covered between 25 and 33 per cent of the R&D expenditures of firms and 25 per cent of the cost of productive investments.[163] The Microprocessor Application Programme, on the other hand, granted £55m. for the diffusion of microelectronics applications and, in particular, microprocessors throughout the industry. This programme was re-funded with an additional £30m. in 1982. In addition, in 1979, the 'Microelectronics in Education' programme was established. The programme, totalling £10m., supported the diffusion of microelectronics applications in schools. Between July 1981 and December 1982, the British government launched a Fibre Optic and Opto-electronics Scheme and a Joint Opto-electronics Scheme (between university and industry), both of which were for five years.

In addition, in 1978 the British government, through the NEB (National Enterprise Board),[164] created INMOS, which specialized in memory devices (initially 16K RAM) and microprocessors. INMOS received £65m. from the government in cash and £35m. in loan guarantees as of May 1984.

British policy, at least after 1978, also supported foreign firms. This policy was neither dependent on the 'domestic characteristics' of foreign subsidiaries (as were French and German policies) nor conditional on joint ventures of foreign producers with domestic firms (as in the French case). British policy supported productive investments, preferably in depressed regions, and R&D programmes of foreign subsidiaries, in specific advanced technological areas. Some American semiconductor producers received support from the British government through these and other programmes. National Semiconductor, for example, received £5m. thorough the Microelectronics Industry Support Programme and additional funds from the Development Planning Agency because it chose to open an integrated circuit plant in Scotland. Fairchild and GEC (British) obtained £7m. from the government (in 1978), following the constitution of a joint venture for producing memories.[165] Finally, ITT received £2m. for producing memory devices in Britain.[166]

In 1983 and 1984, the British government launched two major long-term plans for the support of R&D in the domestic semiconductor industry and of collaborative basic research. The first programme, the Microelectronics Industry Support Programme II (MISP II) was launched in 1984. It relates to the period 1984-90, has a total fund of £120m. and supports up to 25 per cent of the R&D costs of projects in microelectronics. The MISP II is funded by the Department of Industry. Its objectives are the support of the development of custom and semi-custom integrated circuits, and the encouragement of their utilization among users.[167]

The second programme, the so-called Alvey Programme, supports long-term and collaborative basic research in four technological areas, among which are Very Large-Scale Integration (VLSI) and Computer–Aided Design (CAD). It relates to the period 1983-8, has a total fund of £100m. for VLSI and of £30m. for CAD and supports up to 100 per cent of the cost of academic research; up to 90 per cent of industry's basic research of wide interest and up to 50 per cent of other types of industry's basic research.

The Alvey Programme is funded by the Department of Industry, the Ministry of Defence and the Science and Engineering Research Council. Its major objectives are the development of one-micron MOS technology and, more generally, the establishment of a high level of basic research in VLSI technology to make the British industry internationally competitive by the end of the 1980s.[168]

5.4 *Italy*

Italy has not had an effective government policy of support for the semiconductor industry: although an Electronics Plan was passed as part of a larger programme for industrial restructuring in 1978, most aspects of this plan were never implemented.[169]

As a result, domestic firms (i.e. SGS-ATES) received money in the form of R&D support and investment subsidies, but they remained free to pursue their own technological and productive objectives. It is possible, therefore, to say that Italian government policy used to the market mechanism, although it did it in a different way from the German policy.

During the first half of the 1970s, the Italian government supported the domestic semiconductor industry through R&D funding and investment subsidies and through structural intervention in the industry. The R&D funding and investment subsidies were channelled through two general programmes: the 'Applied Research Fund' and the 'Electronics Fund'. The 'Applied Research Fund' was established in 1968 and was directed towards the support of applied research in general (not only in the electronics industry). In 1975 a 'Special Electronics Fund' was established. It allocated L.60bn. ($91.9m.) to support applied research in telecommunications and computers. The fund covered a maximum of 20 per cent of the R&D cost of the projects of firms and 50 per cent of the cost of projects of non-profit institutions and of projects related to international co-operation agreements (Amato, 1983). Between 1970 and 1978, these funds gave SGS-ATES L.998m. ($1.5m.) in R&D support and L7832.4m. ($12m.) in low-interest loans.[170] These funds went into R&D support in power transistors, linear integrated circuits, digital integrated circuits, and particularly MOD technology.

During the first part of the 1970s, STET policy affected the evolution of SGS-ATES. In 1969 IRI-STET developed an electronics plan which established STET as the holding company for IRI (the state-owned Istituto per la Ricostruzione Industriale) in the electronics industry.[171] STET wanted to improve the co-ordination of the R&D activities of ATES (STET) and SGS (Olivetti) in electronic components and intended to build a new plant for active electronics components in central Italy. However, the STET plan was never carried out.

In 1971 STET purchased 60 per cent of SGS[172] and in 1972 STET merged SGS with ATES, the other semiconductor producer of the STET group. The decision to purchase SGS was not part of a coherent strategy on behalf of STET: the STET Plan of 1971–5 made no mention of an intention to purchase SGS.[173] On the contrary, this decision was taken under pressure of external circumstances: in fact, STET followed a policy of 'rescuing' the electronics firm (SGS).[174] Because STET was a public holding, SGS-ATES often had to follow certain public policy objectives. On the other hand, SGS-ATES received a continuous flow of financial funds from the telephone companies within STET.[175] Once STET took control of SGS and merged it with ATES, STET had to restructure and concentrate the former production activities of SGS and ATES. The IRI-STET Electronics Plan of 1974–8 aimed at shifting the concentration of production activities of SGS-ATES away from the consumer market and towards the computer, telecommunication and industrial markets. In 1974–5, however, STET underwent a major crisis which affected the planned expansion of telephone services in Italy and which caused STET to make cuts in all of its programmes, including the one regarding SGS-ATES. As a consequence, SGS-ATES was unable to make such a shift.

In 1977 the Industrial Restructuring, Rationalization and Development Law was approved (Law 675). This law added L.600bn. ($680m.) to the Applied Research Fund: L.400bn. ($453m.) in grants and L.200bn. ($226m.) in other types of support. It also supported up to 40 per cent of the cost of R&D projects and up to 60 per cent of certain R&D projects in some specific sectors (including electronics) in which there was a high degree of industrial risk.[176] In addition, this law provided funds for investments in pilot plants.

Within the framework of this law, a plan for the electronics industry was prepared (1978), but never fully implemented. One part of this plan was devoted to the semiconductor industry; the goals were to maintain the competitive position of Italian industry in linear integrated circuits and power discrete semiconductors and to reduce the technological gap between Italian industry and

and the leading producers of digital integrated circuits, particularly in the LSI-VLSI MOS technology. The plan indicated that support amounting to L.130bn. ($147m.) from 1978 to 1981 was needed for the electronic components industry: L.80bn. ($94m.) in R&D subsidies and L.50bn. ($59m.) in low-interest loans. In addition, the plan stated that the rationalization of the electronic components industry would require between L.100bn. and L.125bn. ($147m.).[177]

Although this plan was never fully implemented, between 1978 and 1983 SGS-ATES received L.62.95bn. ($74m.) in subsidies and L.21.4bn. ($25m.) in low-interest loans for four R&D projects. These projects included MOS digital integrated circuits, linear integrated circuits, power discrete semiconductors, and microcomputer systems.[178]

In the early 1980s, the Italian government approved two programmes of support, one at a general level supporting innovation in industry (1982), and one at a more specific level, supporting R&D in microelectronics (1983). The first of these programmes supported the developments of new products, the application of new processes and the establishment of pilot plants. Electronics was one of the five sectors targeted for support.[179] The second programme was planned to support research in VLSI technology and compound semiconductors in the period 1984-8.[180] The support was planned to amount to L.135.6bn. ($77m.) for research in VLSI technology and L.31.8bn. ($18m.) for research in compound semiconductors.

6 The evolution of the American and Japanese industries

6.1 *The United States: world leader*

During the LSI period, the American semiconductor industry maintained a highly innovative record: most of the major innovations (such as the microprocessor) as well as the incremental innovations and product and process improvements that occurred during this period were, in fact, introduced by the American industry (Braun and MacDonald, 1982). Given the increasing complexity, cost and appropriability of semiconductor technology, the American industry continued to diffuse technological knowledge rapidly and at a reduced cost through labour mobility.

During the 1970s, few merchant producers entered the standard LSI market, while several entered the custom and semi-custom market. The new merchant producers of standard devices were either established very early in the LSI period (around the years 1971-2)[181] or else focused their activites on specific products.[182] The new producers of custom and semi-custom devices concentrated on CMOS and NMOS technologies.[183]

During the LSI period, two major strategic groups continued to compose the American semiconductor industry: merchant producers and vertically integrated producers. Merchant producers occupied a large share of the standard product market.[184] Vertically integrated producers, on the other hand, grew in importance during the 1970s. In 1978 these producers accounted for 20 per cent of world production of integrated circuits and 41 per cent of American production; in 1981 these percentages increased respectively to 23 and 44 per cent (Borrus, Millstein and Zysman, 1982).

As previously mentioned, the technological and productive advantages of American producers in LSI production were translated into exports of LSI devices to the European market, into the successful performance of American subsidiaries in Europe and into the establishment of new subsidiaries in Europe.[185] Intel, for example, continued to supply the European market through exports and established second-source agreements for its microprocessors with several European producers such as Siemens. Texas Instruments, Motorola and National Semiconductor continued to operate in Europe through their European subsidiaries. Texas Instruments had plants in Britain, France and West Germany; these plants specialized in power transitors, MOS integrated circuits and bipolar integrated circuits. Motorola had plants in France, Britain and West Germany; these plants specialized in discrete devices and linear integrated circuits. In addition, Motorola had second-source agreements with EFCIS (France) for the production of Motorola's microprocessors, and technological assistance agreements with Thomson-CSF (France) for the production of bipolar linear devices.[186] Finally, National Semiconductor had a major plant in Britain and in the period 1978–83 controlled 49 per cent of the stock of Eurotechnique.[187]

It is important to remember that two other American vertically integrated semiconductor producers, IBM and ITT, also produced semiconductors on a large scale in Europe. IBM-WTC produced semiconductors for in-house use only: integrated circuits in France and memories in West Germany (Bakis, 1977). ITT Semiconductors in Europe,[188] on the other hand, produced semiconductors both for in-house use and for the external market. ITT also maintained a strict division of labour among its subsidiaries. Intermetall (West Germany) mainly produced discrete devices, MOS and bipolar integrated circuits for the consumer market;[189] Footscray (Britain) produced memory devices for the computer market[190] and Colmar (France) produced diodes. Intermetall was the major ITT producer in Europe, with DM250m. ($138m.) of sales in 1980. Seventy-five per cent of its sales of integrated circuits consisted of custom and application specific integrated circuits.[191]

During the late 1970s and early 1980s, however, two major changes affected the American semiconductor industry: the acquisition of American merchant producers by cash-rich European firms and the penetration of the American market by imports of Japanese LSI devices. The first phenomenon, already analysed, was a result of the development of LSI technology; the second was a result of the improved competitive performance of Japanese producers in specific product ranges, particularly memories. Japanese firms captured 12 per cent of the 4K RAM world market during the second half of the 1970s because they found American firms short of capacity when demand for that type of memory boomed. They also captured 40 per cent of the 16K RAM world market in the later 1970s and early 1980s, and they held 70 per cent of the 64K RAM market in 1981 (see Table 6.8). In the market for 256K RAM, Japanese firms hold 90 per cent of the market (1984)[192] and firms such as NEC, Hitachi and Fujitsu are at the forefront. However, Japanese producers have not penetrated other segments of the American market, where American producers remain the leaders: this is the case, for example, with the microprocessor market.[193]

During the early 1970s, the American semiconductor industry also experienced a wave of downstream integration from LSI device production to electronic consumer goods production such as watches and calculators. Producers such as Fairchild, Texas Instruments and National Semiconductor, for example, integrated downstream into the production of electronic calculators and watches.[194] Since a relevant share of the cost of producing consumer goods came from the cost of LSI devices and since these types of consumer goods had a high elasticity of demand, these semiconductor producers decided to appropriate the value added of downstream consumer goods production.[195]

However, the movement towards downstream vertical integration in the American industry reversed its course during the late 1970s and early 1980s. In that period, in fact, the specific advantages of semiconductor producers disappeared as LSI devices reached the technological level required by watches and other electronic consumer goods. From this moment on, the LSI devices used in watches and other electronic consumer goods became relatively conventional and mature and were available from a large number of semiconductor producers. As a consequence, profits deriving from the introduction of new LSI devices in consumer goods declined and several LSI producers exited from the market.[196]

The convergence of electronics final markets affected the structure of the American industry in a way similar to the European case because it increased the demand for LSI semiconductor devices, and provoked a wave of upstream integration by electronics final goods producers. Several electronics final goods producers did not have experience in integrated circuit production and purchased established semiconductor producers. For example, United Technologies bought Mostek in 1978, and Gould bought American Microsystem in 1981. Among these producers were vertically integrated receiving tube producers that had exited from the standard integrated circuit market during the late 1960s and early 1970s, For example, as Table 6.29 shows, General Electric purchased Intersil in 1981.[197]

Unlike Europe, however, in the United States the convergence of electronics final markets took place in the presence of a competitive computer industry and a competitive semiconductor industry. Consequently, the trend of computer

Table 6.29 Corporate investments in American semiconductor companies by American corporations

Semiconductor producer	Acquiring firm
Analog Devices	Standard Oil of Indiana
Intersil	General Electric
Monolithic Memories	Northern Telecom
Mostek	United Technologies
SEMI	GTE
Spectronics	Honeywell
Synertek	Honeywell
Western Digital	Emerson
Zilog	Exxon
Siliconix	Westinghouse
American Microsystems Inc.	Gould
Intel	IBM

Sources: Borrus, Millstein and Zysman (1982), p. 34; Braun and MacDonald (1982), p. 176; UN (1983), p. 265.

manufacturers to integrate vertically into LSI production (begun in the 1960s) continued throughout the 1970s. American computer producers were highly competitive and profitable; as a result, they could afford to support the temporary losses which were necessary in the short run to gain experience in LSI devices. In addition, because these firms were large (see Table 6.24), they could absorb internally most, or even all, of the in-house production of LSI devices; and because they were technologically advanced in computer production, they could exploit the in-house user-producer interaction. Finally, because most American LSI producers were highly competitive, American computer producers could begin the in-house production of LSI devices either through the acquisition of another American producer or, because of the high mobility of technicians and engineers, through a new venture.

In several cases, the linkages between the semiconductor industry and the computer industry were in the form of co-operation in R&D between semiconductor and computer producers. For example, IBM increased its ties with Intel[198] and co-operated with Texas Instruments and Control Data, while NCR and Digital Equipment co-operated with National Semiconductor and Motorola.[199]

In the United States, the changes in telecommunications and consumer demand for semiconductor devices affected an electronics market that was different from that in Europe. In telecommunications equipment production, ATT-Western Electric had already been present in LSI devices production. In the early 1980s, after the antitrust settlement which divested ATT of its local operating companies, ATT increased its commitment to LSI devices because it

was allowed to enter new markets, such as information processing.[200] In electronic goods production, the penetration of Japanese producers was extensive
in certain product ranges. In the TV market, for example, Japanese producers
began to supply the American market with private-label products, small TV
sets and, later, brand-label products. Japanese producers were favoured by the
fact that the United States had the same transmission standard (NTSC) as Japan.
The successful performance of Japanese producers was also related to the speed
at which they applied integrated circuit technology to TV sets; however, this
speed was accelerated by government support programmes. In 1980 RCA,
Zenith and General Electric had approximately 49 per cent of the American
colour TV market; Matsushita (Quasar-Panasonic) and Sony, 13.5 per cent,
other Japanese brand product producers, 14.7 per cent; Philips (Magnavox,
Sylvania and Philco), 12.2 per cent.[201]

During the LSI period, public procurement was lower than in the 1960s and
did not exert a major pull on innovation as it had before. In 1960, for example,
public procurement in the United States amounted to $258m. (48 per cent of
total shipments), while in 1971 it amounted to $193m. (13 per cent of total
shipments) and in 1973 to $201m. (6 per cent) (Levin, 1982).[202] In addition,
during the LSI periods, the requirement of government markets (particularly
the military) were not the same as those of the civilian market (as they had
been during earlier periods). Whereas, during the 1950s and 1960s both civilian
and military markets needed smaller, more integrated and more reliable devices,
during the 1970s, government markets needed high performance devices which
were suited to specific environmental conditions. These types of devices were
not of any particular interest to civilian markets.

As with public procurement, other public policy factors did not play a major
role in the evolution of the American semiconductor industry either. American
policy towards the semiconductor industry was limited in size: no large programme for the industry was established before 1978. Moreover, government
R&D support was only given in small amounts to a number of firms for projects
in different technological areas; it focused on high-speed devices and on devices
which were highly resistant to high temperatures and radiation.[203] In addition,
the government funded basic research in semiconductor materials, such as
gallium arsenide and III–V compounds (Levin,1982).

The only programme which had some impact on the industry was the VHSIC
Program. This was a seven-year programme established in 1978 which allocated
$200m. to R&D support. The programme granted six groups of (mainly ver-
tically integrated) producers approximately $130m. for R&D in submicron
integrated circuits technology, and approximately $80m. to small-scale R&D
projects in lithography, CAD, software and testing (Levin, 1982). The VHSIC
Program had the following features: first, in contrast with most European
programmes, it was not initiated as a reaction to the commercial pressure (in
the American case, from semiconductor producers to protect them from Japanese

competition). Rather, it originated from the military's need to create a system composed of high-speed integrated circuits. Second, the main goals of this programme (high speed and miniaturization) were similar to the goals of semiconductor producers themselves. Third, this programme focused on the military applications of submicron technology and on complementary areas, such as CAD and software, as well as on the manufacturing aspects of various parts of a more general weapons system. Fourth, like many American military programmes, the VHSIC Program intended to give funds to small and medium-size firms, as well as large, vertically integrated contractors (Texas Instruments, IBM, Rockwell, Raytheon; the group constituted by TRW, Motorola and Univac and the group constituted by Westinghouse and National Semiconductor) (Fast, 1980). The programme had difficulty in fulfilling this objective, however, since very few proposals for R&D projects were submitted by small and medium-size firms.

6.2. *Japan: the successful catch-up story*

The Japanese semiconductor industry is an example of a successful catch-up story. Like the European semiconductor industry, in the early 1970s the Japanese industry was excluded by the bipolar integrated circuit market, dominated by American producers. Yet, unlike the European industry today, ten years later the Japanese semiconductor industry has become a world leader in some LSI devices, such as memories and MOS integrated circuits.

The successful catch-up of the Japanese semiconductor industry *vis-à-vis* the American industry during the LSI period can be explained by supply, demand and policy factors. Early in the LSI period, Japanese vertically integrated producers committed themselves to LSI technology—in particular to MOS integrated circuits. Demand factors, and in particular calculator and computer demand, induced early production of LSI devices. Public policy factors—and in particular government programmes in favour of the computer and semiconductor industries, government co-ordination of firms' R&D, and government protection of the domestic market—supported the efforts of forms in LSI R&D and production. Supply, demand and public factors are briefly examined in the following pages.

During the LSI period, the Japanese semiconductor industry continued to be constituted by vertically integrated corporations.[204] Nippon Electric (NEC), Hitachi, Fujitsu, Toshiba and Oki were among the major Japanese producers of computers and telecommunication equipment; Matsushita, Sanyo, Sony and Sharp, on the other hand, were among the major producers of electronic consumer goods.[205]

Unlike European firms, early in the 1970s Japanese vertically integrated firms committed themselves to LSI technology and to MOS integrated circuits.[206]

Although, like European firms, they were excluded from the digital bipolar integrated circuit market by American firms (such as Texas Instruments) in the late 1960s and early 1970s, Japanese firms began developing MOS integrated circuits for calculators in the second half of the 1960s. Firms such as NEC and Hitachi became convinced of the quality, reliability and potential of MOS technology.[207] These beliefs proved right. MOS technology would, *de facto*, prove the winning technology of the early 1980s. In the early 1970s, several Japanese firms committed considerable resources to the development and production of MOS integrated circuits.[208] NEC and Hitachi were the major producers, followed by Toshiba, Fujitsu, Mitsubishi, Oki and Sharp. During the years 1974–8, Japanese producers rapidly increased their MOS integrated circuit capability. They began exporting integrated circuits in 1974–5 and penetrated significantly into the American memory market in 1977–8.[209] By 1979 the United States had a trade deficit with Japan in integrated circuits, and in MOS technology.[210] In the early 1980s, Japanese producers had captured 70 per cent of the 64K RAM (1982) and 90 per cent of the 256K RAM market (1984).[211] NEC, Sharp, Hitachi and Toshiba were among the major Japanese producers of MOS integrated circuits.

The early building up of technological and productive capability in MOS technology by Japanese producers was complemented by co-operation in R&D and specialization in production. The government encouraged co-operation in R&D among Japanese firms. Co-operation in R&D covered the basic and applied end of the R&D spectrum and was focused on the development of technologies with commercial applications.[212] At the productive level, however, Japanese firms specialized in the production of well-defined products. For example, in the late 1970s, NEC and Hitachi specialized in NMOS and CMOS LSI devices; Fujitsu in NMOS memory and bipolar logic; Toshiba in CMOS LSI devices; and Mitsubishi and Matsushita in linear integraged circuits and some types of memory devices (Borrus, Millstein and Zysman, 1982, p. 66). At the firm level, firm co-operation in R&D and government policy reduced the cost and lowered the appropriability of technological knowledge, while continuing to maintain a high appropriability of the development end of the R&D spectrum and of product and process innovations. While co-operation in R&D avoided duplication of effort, brought together firms with different technological capabilities and reduced the cost of doing R&D in an increasingly complex technology such as the LSI technology, specialization in production allowed the exploitation of static and dynamic economies of scale.[213]

Among demand factors, calculator and computer demand for semiconductors played an important role in the successful performance of the Japanese industry in MOS technology and memory devices. The calculator market was responsible for the early commencement of Japanese production of MOS integrated circuits. Japanese producers had been extremely successful in calculators. In 1971, for example, Japanese firms had approximately 85 per cent of the American calculator market. This share declined over the 1970s, but always remained relatively high.[214] In the early periods of calculator production, Japanese producers purchased integrated circuits from Rockwell, GI, Texas Instruments and Fairchild.[215] However, following the beginning of NEC production of MOS integrated circuits for desk-top calculators in 1966,[216] in the early 1970s Japanese semiconductor firms began producing MOS integrated circuits to meet domestic calculator demand. Several of these producers were vertically integrated into calculator production. In 1976 the calculator demand for integrated circuits still accounted for 14 per cent of the total Japanese demand for integrated circuits. This percentage decreased to 8 per cent in 1978.[217]

Computer demand stimulated domestic production of LSI devices later in the 1970s. The Japanese computer industry grew dramatically, and was relatively successful during the LSI period. Japanese firms held a share of the Japanese market amounting to 48 per cent in 1967, 58 per cent in 1976 and 53 per cent in 1983.[218] The major Japanese computer firms were Fujitsu, Hitachi and NEC. These firms were also vertically integrated into the production of LSI memory devices. It should also be noted that during the 1970s the Japanese computer industry continued to be supported by public policy, which protected the home market, followed a policy of 'buy Japanese' and launched several support programmes.

During the LSI period public policy factors played an important role in the development of Japanese LSI production. Japanese public policy was formulated early in the LSI period, was coherent and consistent, and co-ordinated rather than directed domestic firms. In the early 1970s, MITI targeted the semiconductor industry as a strategic industry and as a prime object for support. From 1972 to 1977, the Japanese government increased support for the R&D of domestic firms as a compensation for the partial opening of the domestic market.[219] Only in 1976, however, did the Japanese government establish a VLSI Plan. This plan covered the period 1976-9 and went to finance a development project for large-scale integrated circuits that were to be used in a new series of computers; the financing totalled Y30bn. ($120m.) (42 per cent of the total cost) (Fast, 1980, p. 121). The programme supported two research groups: the first contained Mitsubishi, Fujitsu and Hitachi; the second NEC and Toshiba (Peck, 1983). In addition, the Japanese government protected the home market with a quota system until 1974 and with a foreign direct investments bar until 1976.

Conclusions

During the LSI period, the characteristics of semiconductor technology, and the linkages and interdependencies of the semiconductor industry changed.

Semiconductor technology became more complex, cumulative and appropriable. Semiconductor devices performed an increasing number of linear functions in a digital way, and they became more closely integrated with the developments of the computer industry. Finally, a technological convergence of computer, telecommunication and consumer products occurred.

During the first part of the 1970s the situation of the European semiconductor industry was the same as it had been at the beginning of the LSI period. European semiconductor firms continued to produce discrete devices, linear integrated circuits and speciality and custom digital integrated circuits, while American firms dominated the LSI devices market.

It took some time before the new technology, linkages and interdependencies had their effect on the European semiconductor industry. Not until the late 1970s, in fact, did the increasing digitalization of linear functions and the convergence of the electronics final markets threaten the established competitiveness of the European vertically integrated producers in linear integrated circuits, consumer electronics and telecommunication equipment. In order to survive in the changing semiconductor and electronics industries, these vertically integrated producers had to enter and commit themselves to LSI devices production. This was the case with regard to Siemens and Philips, for example.

However, because of the complexity, cumulativeness and appropriability of LSI technology, these firms were not immediately successful in their entry and commitment. Only with time, in fact, could European firms build up technological capabilities and accumulate experience in LSI technology. In order to improve their possibilities in the LSI market, these firms purchased American LSI merchant producers and, later on, began R&D collaborations with other producers of LSI devices.

Unlike the European industry, the Japanese industry was increasingly successful during the LSI period. The demand structure in Japan (a strong calculator demand in the late 1960s and an increasingly competitive computer industry during the 1970s) provoked an early commitment of vertically integrated Japanese producers to LSI technology—in particular to MOS technology.

The comparison of the European case with the Japanese emphasizes how subtle the relationship is between competitiveness at the final product level and at the input level, particularly for vertically integrated firms. If the focus of the firm is on the final product, then the purchase of non-state-of-the-art components from in-house sources or from domestic producers cannot be sustained for a long time without impairing the competitiveness of the final product. On the other hand, if the in-house sources or the domestic producers have advanced technological capabilities, a highly innovative or internationally competitive final goods firm may not only receive innovative inputs from these producers, but also provide them with an innovative stimulus and a useful feedback.

This type of relationship between final goods and component producers explains why several European government policies in favour of the domestic computer industries failed to affect the domestic semiconductor industry and why a large-scale European production of LSI devices could only begin within vertically integrated producers. Non-vertically integrated electronics final goods producers had to continue to buy state-of-the-art LSI devices from American producers in order to survive. They could not subsidize the growth and experience of newly established semiconductor producers. Vertically integrated producers and the government, on the other hand, could: EFCIS and INMOS are cases in point.

While the importance of public policy declined during the LSI period in the American industry, it increased in the European and Japanese industries. However, while Japanese policy continued to protect the domestic semiconductor market for most of the 1970s, was coherent and consistent through these years and co-ordinated rather than directed domestic firms, European policies differed widely in timing, scope and tools. Most European policies were late compared to Japanese policy, and were less coherent and more directive. Only the German policy had timing, a rather coherent framework and a less directive and more market-orientated approach. While Japanese policy was successful in improving the international competitiveness of the domestic industry, most European policies were not. Only the German policy was partially successful.

Differences in public policy and in the R&D co-operation among firms between Japan and Europe had major consequences for the appropriability conditions of these countries.[220] Given the increasing complexity and cost of semiconductor technology, the diffusion of technological knowledge among firms was costly and required time. In Japan, R&D co-operation among firms and government co-ordination of the R&D activity of firms increased the rate of diffusion of technological knowledge among firms, without consistently reducing the appropriability of the returns of innovation at the firm level. On the other hand, the reduced co-operation among European firms in R&D meant that firms maintained a high appropriability of technological knowledge.

It is interesting to note that, while the co-operation among firms and public policies in Japan acted on appropriability conditions through the diffusion of technological knowledge at reduced costs, in the American semiconductor industry a different factor acted in the same direction: labour mobility. The mobility of personnel among semiconductor producers, in fact, diffused technological knowledge in the American industry through its influence on the technological capabilities of firms. In this case, as well, the low degree of labour mobility among European firms impeded the working of this mode of diffusion of technological knowledge.

By the early 1980s, several European vertically integrated producers were active in the LSI market. Philips and SGS-ATES produced mainly custom and application-specific integrated circuits. After major restructuring, Thomson-

CSF attempted to produce standard and custom LSI devices. Ferranti and Plessey, on the other hand, specialized in semi-custom LSI services. Most of these producers had become profitable in recent years. Only Siemens and INMOS mounted a major entry in the standard LSI market, and particularly in memories and microprocessors.

Notes

1. Among the numerous introductory reports on this topic, see *Scientific American* (1977).
2. These devices combine the best properties of the two transistors. They were introduced by National Semiconductor in 1970, and by RCA in 1972. Source: Wilson, Ashton and Egan (1980).
3. The I2L is a bipolar integrated circuit and derives from the DEL. It has a very high packing density and high speed. It was introduced by Philips in 1975.
4. The VMOS is able to handle high frequencies. It is used in high-power MOS and ROM. The VMOS was introduced by Siliconix and AMI in 1978. Sources: Young (1979), Wilson, Ashton and Egan (1980).
5. The main process innovations included sputtering; polisilicon deposition, introduced by Intel in 1973 (Wilson, Ashton and Egan, 1980); silicon-on sapphire (SOS) used in MOS fabrication for increasing the speed of the circuit, introduced by RCA in 1973 (Wilson, Ashton and Egan, 1980); the locally oxidized CMOS (LOCMOS) for high performance, high density CMOS, introduced by Philips in 1975 (Dummer, 1978); the H-MOS (high-performance MOS) introduced by Intel in 1977 (Dummer, 1978); the co-planar (isoplanar) used for reducing the size of the transistor, introduced by Fairchild in 1976 (Young, 1979; Streetman, 1980; Wilson, Ashton and Egan, 1980); the plasma etching, introduced by STL (ITT) in 1976; the ion milling, introduced by Veeco in 1975 (Wilson, Ashton and Egan, 1980).
6. Perkin Elmer and Bell Laboratories were developing X-ray lithography in the early 1980s. The electron beam processing was introduced by Bell, Texas Instruments and IBM in 1975 while photolithography projection alignment was introduced by Perkin Elmer in 1976 and deep ultraviolet photolithography was introduced by Perkin Elmer in 1977 (Braun and MacDonald, 1982; Wilson, Ashton and Egan, 1980).
7. This occurred particularly after the introduction of the microprocessor; the interfaces transformed an analog signal into a digital signal, then transmitted, transformed, and eventually reconverted it into analog signals. Digital signals can, in fact, be transmitted and represented more accurately than analog signals (Hazewindus, 1982, p. 73).
8. For a discussion, Braun and MacDonald (1982); Harewindus (1982); Ernst (1981).

9. However, this technology remained difficult to master and the dissipation of these transistors was high (Hazewindus, 1982).
10. According to Braun and MacDonald (1982), optical memories have the advantage of being permanent, of having high density and of being compatible with optical fibres. See also Young (1979).
11. Magnetic bubble memories can store the information when the voltage is switched off (Braun and MacDonald, 1982; Hazewindus, 1982).
12. CMOS became widely used when solutions to their slowness and bulk were found, so that their advantages in terms of low power consumption and easy design could be better exploited. CMOS 32-bit microprocessors and CMOS dynamic RAM have been developed by Intel and Motorola, and by Intel, respectively (*Wall Street Journal*, 22 February 1983).
13. Compound rates. Dataquest in United Nations (1983), p. 15.
14. US Department of Commerce (1979) and Webbink (1977).
15. See the publications of ZVEI (West Germany), various years.
16. See the publications Sycep-Sitelesc (France), various years.
17. Dataquest in *Business Week* (28 June 1982).
18. *The Economist*, 8 November 1980 and 18 December 1982.
19. For a complete list of the second source agreements, see United Nations (1983).
20. For a discussion, see Finan (1975) and US Department of Commerce (1979).
21. In addition, some of the major European corporations, such as Philips and SGS-ATES, had offshore assembly and production facilities. This affected the value of imports in the European countries.
22. Mainly because of TTL production (source: interviews).
23. In the late 1960s, in semiconductor production, as well as in other production lines, Philips experienced an uncontrolled proliferation and a highly decentralized mode of operation. For a general overview, see *Forbes*, 1 September 1972.
24. Source: interviews.
25. This components division was unprofitable during the early 1970s, *The Financial Times*, various issues.
26. Source: Siemens Berichte (various years). As will be discussed later, Siemens' components division also sold components purchased on the external market. In 1975, as a result of the crisis of the electronic components market, this share stabilized at 5 per cent.
27. Olympia produced computers.
28. Source: interviews.
29. Lorenzi and LeBoucher (1979), p. 133. In 1976, with sales of 313m. frs. ($65m.), Sesocosem had losses of 100m. frs. ($21m.).
30. Sciberras (1977) divides semiconductor producers in Britain into 'big league' and 'small league' producers. Big league producers (Texas Instruments, Motorola, National Semiconductor, Fairchild) produced standard products for the world market. Small league producers (ITT, Plessey, Ferranti, Lucas) produced custom products for specific users in the domestic market.
31. Source: AGS-ATES.

32. In 1971 ATES had L.2 bn. ($3.2m.) of social capital, while SGS had L.6.5bn. ($10.5m.) of social capital.

33. During the early 1970s, SGS-ATES also produced MOS integrated circuits and maintained a good competitive position on the European COSMOS integrated circuits market. Souce: interviews.

34. The STET holding owned several other firms producing electronics products: Sit-Siemens (telecommunications equipment); Selenia (telecommunication equipment, military and industrial equipment); Elsag (military and industrial equipment). These firms had large shares in specific Italian electronics final markets: for example, in 1971–2, Sit-Siemens and Selenia had 37 per cent of the Italian telecommunication equipment market (Antonelli and Lamborghini, 1978, p. 80).

35. In general, STET did not closely monitor its firms. STET's firms were, to a certain extent, free to choose different productive strategies.

36. These lines were also produced by Signetics. Source: interviews.

37. In 1980 Signetics jumped to sixth place, Levin (1982), p. 238.

38. Sciberras (1977) and interviews.

39. Source: interviews.

40. Source: interviews.

41. Source: interviews. Mullard (GB) developed and produced these devices. The Prestel system is used by British Telecom. Not all of Philips' developments in this field proved successful. For example, Philips' frequence synthetizer and digital tuning integrated circuits proved unsuccessful.

42. Even though patents are not a major competitive tool in the semiconductor industry (see Chapter 2), they provide useful indications of the technological specialization of firms.

43. Source: interviews.

44. *The Economist*, 28 November 1981.

45. *Elektronik-Anzeiger* (October 1978) and interviews.

46. The largest customer for Siemens' production of memories and microprocessors was Siemens' computer division. *Elektronik*, April 1978 and interviews.

47. *Business Week*, 13 October 1980, and interviews.

48. *Wirtschaftswoche*, 26 November 1979, and interviews.

49. *Elektronik*, 22/77.

50. *Forbes*, 8 December 1980 and SEC Forms 10K (various years). AMD did not produce semiconductor devices for the consumer market.

51. A firm specialized in discrete semiconductor devices. *Wirtschaftswoche*, 26 November 1979.

52. An American firm specialized in liquid electronic displays. *Elektronik* (1977 H7).

53. During the late 1970s, Siemens' shares of AMD, Litronix and AMC were respectively 20, 100 and 94 per cent. Source: SEC Forms 10K, various years, and *Business Week*, 13 October 1980.

54. *Elektronik-Applikation*, 13 September 1981.

55. *Wirtschaftswoche*, 26 November 1979; *Elektronik* (1979 Hll).

56. *Elektronik Applikation*, November 1980.

57. *Elektronik Applikation*, 13 May 1981.
58. *Business Week*, 28 November 1981. In its overall activities, Siemens remained profitable during the 1970s.
59. Siemens Components, February 1980, and interviews.
60. Approximately 50 per cent. Source: interviews.
61. Of the type 8080/8085/8048/8086.
62. *The Economist*, 28 November 1981 and *Elektronik* (1980 H21). Siemens started to produce 64K RAM in 1982.
63. *Business Week*, 28 November 1981; *Elektronik Applikation*, November, 1980, p. 12; *Business Week* 13 October 1980. The in-house use of semiconductor production of Siemens' Component Division is different from the in-house sales of Siemens' Component Division, because the Division purchased semiconductors in the open market and sold them to other divisions in Siemens, thus behaving *de facto* as a purchaser of components for the whole corporation. Thus, in 1980, the semiconductor sales of Siemens' Component Division consisted of two-thirds of its own production and one-third of external production; the sales of microprocessors and memories consisted of 50 per cent of its own production (*Elektronik*, 21/80). For microprocessors and memories, Siemens' Component Division planned to increase the ratio of internal production/external production from 1:1 to 3:1 (*Elektronik Applikation*, 11/80). The percentage of electronics components sales going into in-house uses was approximately equal to 30 per cent in the mid-1970s; it increased steadily during the second half of the 1970s.
64. Power MOSFETs have high-frequency performances that are superior to bipolar power transistors and are well suited for switching power supplies. In addition, Siemens introduced surface-wave filters and glass fibre components for telecommunication equipment.
65. *Elektronik Applikation*, November 1980, inverviews.
66. AEG-Telefunken Berichte (1981).
67. *Elektronik Applikation*, November 1980.
68. This share was greater than 10 per cent in the early 1970s. Source: interviews.
69. Source: interviews.
70. Thomson-CSF's semiconductor division was formed by Sescosem and le Silicum Semiconducteur, a producer of discrete devices. Thomson-CSF was a world leader in discrete devices (power transistors). It relied on Motorola's technology in its production of bipolar integrated circuits. Integrated circuits constituted a smaller part of Thomson-CSF's semiconductor sales than discrete devices.
71. Atomic Energy Commission.
72. Source: interviews and EFCIS.
73. Source: interviews and EFCIS.
74. Only about one-third of standard devices production went to the internal market.
75. Source: interviews.
76. *Business Week*, 16 February 1980, 11 August 1980, and interviews.

77. Source: interviews.
78. Source: interviews.
79. From slightly over 10 per cent in 1980 to approximately 20 per cent in 1984. Source: interviews.
80. Source: interviews.
81. Source: Ferranti and interviews.
82. Source: interviews.
83. 16K static RAM, at the beginning, and 64K RAM and the transputer (memory plus microprocessor) later. *The Economist*, 28 November 1981; *Electronics Times*, various issues, and interviews.
84. *The Economist*, 29 October 1983; *Electronics Times*, various issues, and interviews.
85. In the early 1980s, Mullard (Philips), and various American firms, such as Texas Instruments, Motorola, NSC, Fairchild and Intel, continued to control a large share of the British market, either through their plants in Britain or through imports. Mullard, for example, produced linear integrated circuits for consumer applications, teletext type circuits for consumer-telecommunications applications, ROMs and static RAMs, and NMOS integrated circuits. In addition, in the late 1970s and early 1980s the clustering of new British electronics firms increased along the M4 motorway, west of London. This process had begun in the early 1960s. However, most of the firms involved were (and remained) small (*The Economist*, 11 July 1981 and 30 January 1982).
86. In 1979 discrete semiconductor devices accounted for 35 per cent of SGS-ATES sales, and linear integrated circuits for 35 per cent, while digital integrated circuits accounted for 15 per cent (interviews and SGS-ATES). These products were sold mainly outside the STET group to Olivetti, and other consumer electronics producers (i.e. Grundig); Sit-Siemens (now Italtel) and Selenia were the major buyers within the STET group. None of these buyers, however, ever bought more than 10 per cent of the products of SGS-ATES. Olivetti mainly purchased discrete, power and linear devices for its typewriters.
87. STET telecommunication equipment producers.
88. See Lamborghini and Antonelli (1981). In 1980 Olivetti absorbed SGS-ATES's discrete, power and linear devices, and 50 per cent of SGS-ATES's production of Z80 microprocessors, second-sourced from Zilog regarding this product, Olivetti satisfied 80 per cent of its Z80 requirements with purchases from SGS-ATES and 20 per cent with purchases from Zilog (Mazzoni-*L'Elettronica*, 15 April 1981).
89. Source: interviews.
90. Particularly the F8—with Fairchild licence—and the Z80 and Z8000—with Zilog licence.
91. Source: interviews and SGS-ATES. In 1980 Z80 microprocessors controlled 33 per cent of the Italian market for microprocessors; Intel 8080s and 8035s controlled 33 per cent; Fairchild F8s 6 per cent; Motorola 6800–68000s 6 per cent. Olivetti constituted 40 per cent of the demand for microprocessors in Italy (Mazzoni-*L'Elettronica*, 15 April 1981).

92. SGS-ATES held, respectively, between 6 and 8 per cent and between 8 and 9 per cent of the European market in the early 1980s in these product areas. In the early 1980s SGS-ATES decided to stop its production of some type of discrete devices and of bipolar digital integrated circuits (source: interviews).

93. As previously seen, in 1983 Eurosil was purchased by Telefunken and United Technologies. In 1977 Bosch purchased the 25 per cent of AMI, an American semiconductor producer.

94. See, for example, Flamm-Pelzman (1984).

95. For an analysis of the dynamic competition of American firms over several national markets, see Flaherty (1984).

96. This incentive for vertical integration will be examined in more detail in the following sections.

97. For a more detailed analysis, see United Nations (1983) and Flamm-Pelzman (1984). As will be examined later, the same type of phenomenon occurred in the United States. American cash-rich electronics final goods producers took control of American merchant producers.

98. This type of collaboration must be kept separate from the two common forms of technology transfer among semiconductor producers: second-source agreements and licence. Second-source agreements refer mainly to product technology, while licences refer mainly to process know-how and technology. During the history of the semiconductor industry the former have been more widely used than the latter.

99. As will be seen later, the Esprit program of the EEC favoured collaboration in R&D among European firms.

100. As will be shown later in the chapter, co-operation in R&D has been successfully used, first in Japan, and more recently in the United States.

101. Infocorp in *Business Week*, 16 July 1984.

102. Minicomputers were introduced in the early 1960s by electrical engineering firms for military applications. Only in 1965, however, was the first general purpose minicomputer, the PDP-8, built by Digital Equipment Corporation (DEC). See Sciberras, Swords-Isherwood and Senker (1978).

103. The only major exception is Nixdorf in the mini and small computer market.

104. During the early 1970s, although its technological agreement with RCA was interrupted by RCA's exit from computer production and although its Unidata venture failed, Siemens was 'determined to stay in computers because Siemens needed computers for its own development. Although it was in loss, Siemens will not stop producing computers' (*The Economist*, 23 August 1975).

105. *The Financial Times*, 20 July 1974.

106. OECD (1977); US Department of Commerce (1979). See also Grant and Shaw (1979); Rösner (1978); Kloten *et al.* (1976) for a detailed analysis of the German computer industry. Philips also had only a limited involvement in computer production, following the failure of the Unidata venture.

107. CII was created by the Plan Calcul in 1966. Honeywell-Bull was created in 1970 when General Electric sold its activity in Bull-GE to Honeywell.

108. *Business Week*, 26 October 1974. Olivetti had sold its computer activities to General Electric in the mid-1960s.
109. In 1970 only 16 per cent of Olivetti's products were electronic; in 1978 this share increased to 55 per cent (Lamborghini-Antonelli in OECD, 1981, p. 28). In the second half of the 1970s, Olivetti also marketed the large and medium-size computers of Hitachi and IPL. Source: interviews.
110. Total units installed (US Department of Commerce, Country Market Survey: West Germany) 1976; Grant and Shaw (1979).
111. Grant and Shaw (1979). Grant and Shaw claim that in 1974 44 per cent of the European public sector demand for computers was satisfied by European firms.
112. US Department of Commerce (1976); Grant and Shaw provide a value of 35 per cent. ICL had a share of 46 per cent of the public sector market (Grant and Shaw 1979).
113. The first programme (1967–70) of $212.63m. supported mainly computer hardware and large computers. See BMFT Reports, various years.
114. Grant and Shaw (1979). In medium and large-size computers, Siemens received DM 200m. ($87m.) up to 1975, DM 60m. ($24m.) in 1976, DM 46m. ($20m.) in 1977 and DM 45m. ($22.5m) in 1978. Source: BMFT (1980).
115. Fusfeld and Tooker (1980); Dosi (1981). The French government also supported the creation of CII-Honeywell-Bull.
116. In 1982, after years of losses, ICL showed some profits. *Business Week*, 20 July 1981 and *The Economist*, 11 December 1982.
117. It is necessary to remember once again, that government intervention allowed the entry of some new European merchant producers. EFCIS and INMOS are the two major examples. EFCIS entered the European custom LSI device market with French government support. Later on, it expanded its production to standard LSI devices. INMOS entered the standard LSI device market with extensive British government support. Government intervention subsidized the first years of losses of these producers.

    The survival of EFCIS and INMOS as merchant producers was short. Within a few years, both EFCIS and INMOS had been purchased by electronic final goods firms. EFCIS became part of Thomson-CSF and INMOS of Thorn.
118. This was the result of the digitalization of transmission, the change from electro-mechanical to electronic devices in switching systems and the convergence of the computer, telecommunication and office equipment products in terminal equipment.
119. In 1982 France invested $6.8bn., West Germany $4.0bn.,, Britain $3.8bn. and Italy $3.4bn., vs. $31.2bn. of the United States and $9.9bn. of Japan (*Mondo Economico*, 1 March 1984).
120. The local source content in West Germany, France, Britain and Italy was between 85 and 95 per cent (OECD, 1981).
121. The Post Office and British Telecom, since 1981.
122. This market increased in importance during the late 1970s and early

1980s. In the late 1970s, the size of the PABX market was approximately $1.5–1.8bn. (OECD, 1981).

123. Most of these European firms were diversified electronics firms. Other producers, such as Grundig, were more specialized producers: Grundig had approximately half of its sales in colour TV receivers.

124. Which resulted in the voluntary restrictions of exports of video cassette recorders by Japanese producers.

125. During the 1970s European procurement policies continued to be limited in size and characterized by a lack of innovation. Although they became more consistent during the 1970s than during the 1960s, they remained of limited importance. During the 1970s European government markets only constituted (approximately) 5–6 per cent of electronics final markets in Europe (see Table 6.22). In West Germany, for example, the amount of electronic components absorbed by the military increased from $50m. in 1975 to $130m. in 1980. This increase (+160 per cent) was higher than the increase in the size of the market for electronic components (+58 per cent) and for semiconductors (+160 per cent) was higher of Commerce—Country Market Survey, 'Electronic Components—Germany, June 1978). In the late 1970s, approximately 8 per cent of the total sales of Siemens and 10 per cent of the total sales of AEG-Telefunken went to the military. In France, public procurement continued to be mainly directed towards the electronics products of Thomson-CSF. In the mid-1970s, for example, Thomson-CSF covered 61 per cent of public procurement of telecommunication, radio and television products and 95 per cent of detection products (Lorenzi and LeBoucher, 1979, p. 154). It has been estimated (*Fortune*, 28 July 1980) that in 1980 the French government, either directly through purchases of semiconductors, or indirectly through purchases of electronic equipment, controlled 40 per cent of the semiconductor market in France.

126. Source: interviews.

127. It is interesting to notice that at the uninitiated level, both *Science* and *Scientific American* in 1977 could predict with a certain accuracy the technological trends in LSI of the next several years.

128. Source: interviews. A Siemens chief executive explained: 'Siemens is not in computers or ICs just to compete with IBM or Intel, but to stay current in technologies crucial to its strongholds in such markets as communications, industrial systems and medical electronics. All told, nearly all of Siemens' entire business is dependent on electronics' in *Business Week*, 1 February 1982, p. 84.

129. For a discussion of appropriability of technological know-how in the American semiconductor industry, see Flaherty (1984). For a discussion on the importance of design and software, see United Nations (1983) and Ernst (1981).

130. The creation of an in-house capability and the policies of acquisitions and of co-operation have been described earlier.

131. Even though several European electronics firms vertically integrated into LSI production, this solution was not the only one available for firms

aiming at obtaining knowledge and mastery of LSI technology. Another
solution was the beginning of R&D in custom LSI devices and their production through silicon foundries or through the adaptation of semi-
custom devices. See Hazewindus (1982) and Braun and MacDonald (1982).
132. Source: interviews.
133. Source: interviews. For a general discussion of German public policy,
see Keck (1976).
134. Source: interviews.
135. In 1972 and 1973, such support amounted to DM4.95m. ($1.5m.) for
R&D in solid-state physics, to DM5.29m. ($1.6m.) for R&D in electronic
components and to DM8.02m. ($2.5m.) for R&D in optoelectronics
devices. Source: BMFT (1974).
136. The official reasons for the programme included the importance (for the
whole electronics industry) of a competitive domestic semiconductor
industry; the need for a stable supply of semiconductor components
for the domestic electronics industry and the high R&D cost and high
risk involved in semiconductor activity (BMFT, 1974).
137. Molecular beam epitaxy, controlled gaseous phase epitaxy, electron
beam exposure, microstructure.
138. X-ray topography, ion-back scattering and secondary mass spectroscopy.
139. The microelectronics support plan was divided into six sections: process
technology; design and systems; peripherals; microelectronics application;
materials and basic research; general projects and studies. As in previous
programs, R&D support covered 50 per cent of the cost of a project in
most cases and 100 per cent in a few special cases.
140. Between 1982 and 1984 these programmes amounted to DM450m.
(approximately $176m.).
141. The electronics funds amounted to DM118.1m. ($51.3m.); the Data
Processing funds to DM145.5m. ($63.2m.); the Communication Techno-
logies to DM52m. ($22.6m.) and the Information Technologies to DM24m.
($10.8m.). In addition, other programmes totalled DM12.2m. ($5.3m.).
These programmes included the Societal Aspects of Information, the Basic
System Service, the funding of institutions, the GMD society for Data
Processing and the Heinrich Hertz Institute. Source: Fusfeld and Tooker
(1980).
142. In addition to these programmes, a synchrotron radiation project was
financed in Berlin. Siemens, Philips, Eurosil and AEG-Telefunken participated in this project. See Fusfeld and Tooker (1980).
143. BMFT (1984) and interviews.
144. For a general discussion, see Suleiman (1975), Zysman (1975 and 1983).
145. For a detailed analysis, see Le Bolloch (1983).
146. France-Commissariat Général du Plan (1976), p. 27.
147. France-Commissariat Général du Plan (1976), p. 35.
148 Called *filières*.
149. Sescosem became part of Thomson-CSF in 1972 in order to hide its losses
(Pottier and Touati, 1981).
150. Collaboration agreements for the development and delivery of these types

of integrated circuits were established between RTC and CII-Honeywell-Bull, *Le Nouvel Economiste*, 26 July 1978.

151. Eurotechnique was a joint venture between Saint Gobain Point-à-Mousson (51 per cent on the capital) and National Semiconductor.
152. MHS was a joint venture between Matra (51 per cent of the capital) and Harris.
153. Sources: Pottier and Touati (1981); *Le Nouvel Economiste*, 26 July 1978; Vol Argent, February 1979; interviews.
154. Thomson-CSF increased its shares in EFCIS from 34 to 68 per cent.
155. Quatrepoint, *Le Monde*, 11 December 1979 and 15 April 1980, Pottier and Touati (1981) and interviews.
156. The funds for all these programmes were channelled through several institutions. The Ministry of Industry provided the largest share in the form of both investment subsidies and R&D support. Other funds came from the PTT, the Ministry of Telecommunication and the Military.
157. Gougeon, Ponson and Tinard (1984) and Truel (1983).
158. In 1982 the Ministry of Industry was supposed to fund almost half of the support for R&D; the Ministry of Defence and the DGT approximately one-fifth and the Ministry of Research the rest. Sources: *Le Nouvel Economiste* (various issues); Quatrepoint, *Le Monde* (various issues); interviews.
159. Source: UK Electronic Components Economic Development Committee (1983) (I. Maddock) and interviews.
160. Source: interviews.
161. Now called Components, Valves and Devices.
162. In 1969 and 1970 GEC's share was greater than 35 per cent of the total number of contracts assigned (Dickson, 1983).
163. The largest share of support went to the development of manufacturing processes and production facilities for custom and semi-custom devices. Ferranti's uncommitted logic arrays were one of the major projects funded. Source: interviews.
164. Later BTG (British Technology Group).
165. Source: interviews. This joint venture never took effect, however.
166. Source: interviews.
167. Source: UK Department of Trade and Industry and interviews.
168. Source: UK Department of Trade and Industry (1983) and interviews.
169. Adams (1985).
170. Source: SGS-ATES.
171. According to the plan (1970–80), STET was to develop its manufacturing and services activities in four areas: electronic components; telecommunication services and equipment; instruments and automation; and computers (IRI-STET Plan, 1969).
172. Olivetti and Telettra each kept a 20 per cent share of SGS until 1975, when they sold the shares to STET.
173. See Antonelli and Lamborghini (1978) for a detailed analysis of the STET policies of these years.
174. According to the STET Plan 1971–5, STET concentrated its investments

in telephone services and in telecommunication equipment production. STET did not seem to pursue the creation of the vertically integrated and diversified structure that was envisaged in the earlier IRI-STET Plan (see Antonelli and Lamborghini, 1978, p. 77).

175. This flow originated from the higher cash flow (profits plus depreciation), social capital and cash-flow/sales in STET's telephone service companies, compared to other STET electronics firms like SGS-ATES. This flow was mainly channelled through inter-company prices and inter-company loans. In fact, from 1973–6, an average of 48 per cent of SGS-ATES's debt was financed by intercompany loans (Antonelli and Lamborghini, 1978, p. 195).

176. The high risk was defined in terms of the uncertainty surrounding the technological evolution of the market; of the presence of a relevant number of foreign competititors; and of the amount of resources involved in the project.

177. The electronics plan introduced two innovative aspects into Italian government policy towards the electronics industry. First, it recommended that public procurement should become a more important part of industrial policy. Before the plan, public procurement was rarely used as an instrument of industrial and/or innovative policy in Italy. Second, the electronics plan made foreign subsidiaries eligible for support on the condition that they establish technologically advanced R&D facilities in Italy and that they balance their imports with their exports.

178. Source: SGS-ATES and interviews.

179. The other sectors were auto, steel, aeronautics and fine chemicals.

180. Italy — Ministry of Scientific and Technological Research (1983).

181. The years 1968–72 in fact represented one of the three periods of peak entry. See Wilson, Ashton and Egan (1980).

182. In this way, they were not at a major disadvantage compared to established LSI producers. For example, Zilog successfully competed in the microprocessor market.

183. For a list, see HTE Management Resources, in *l'Elettronica*, No. 12, 1984.

184. Merchant producers could be divided into subgroups. According to Wilson, Ashton and Egan (1980), firms such as Texas Instruments aimed at major innovations in a wide range of semiconductor products; other firms, such as Intel and Mostek, aimed at major innovations in specialized products; still others, such as Motorola and National Semiconductor focused their strategies on incremental innovations over a wide range of products while others, such as AMD, Intersil, and Zilog focused their strategies on incremental innovations for specialized products.

185. During the 1970s, American firms continued to invest in offshore assembly, in order to profit from low labour costs (Finan, 1975; United Nations, 1983). In the early 1980s, however, because of the automatization of the assembly process, this trend was partially reversed (Ernst, 1981).

186. Source: interviews.

187. Saint-Gobain-Point-à-Mousson was the other partner. This stock was sold to Thomson-SCF in 1983.

188. In the United States, an ITT plant in Florida produced bipolar devices and another plant in Massachusetts produced discrete devices. Source: interviews.

189. Intermetall also second-sourced Texas Instruments' 16-bit MOS microprocessor 9900 for consumer electronics applications, *Elektronik* (1980 H6) and interviews.

190. It produced 4K and 16K dynamic RAMs for the computer market (*Wirtschaftswoche*, 31 March 1978). Footscray is now sold to the British.

191. *Elektronik Applikation* (1980), No. 11, p. 12 and interviews. Other American subsidiaries (General Instruments, Analog Devices and International Rectifier) produced semiconductors in Britain.

192. *The Economist*, 23 February 1980 and 20 March 1982; *The New York Times*, 28 February 1984 and 6 September 1984; *Business Week*, 15 March 1982.

193. *The Economist*, 8 November 1980 and *The New York Times*, 6 September 1984.

194. Texas Instruments, National Semiconductor and Rockwell began to produce pocket calculators in 1973. Texas Instruments, Intel, National Semiconductor and Fairchild began producing electronic watches in 1973–5. In both the pocket calculators and the electronic watches businesses, these firms followed the same aggressive policy used in their semiconductor business (Wilson, Ashton and Egan, 1980; Braun and MacDonald, 1982). Competition among the LSI producers was very keen. In 1978 Fairchild had to exit from the watch business because of the aggressive policy of Texas Instruments and because they had made some wrong technological choices (Wilson, Ashton and Egan, 1980).

195. Provided that they did not have any major cost disadvantage in the production and marketing of consumer goods. By producing standard LSI devices, semiconductor firms faced low transaction costs, and ran no risk of being taken over by the downstream consumer goods producers. By being innovative in LSI devices, these firms enjoyed temporary monopoly. This downstream incentive to vertically integrate can be referred to as the incentive of a monopolist to integrate downstream into a competitive industry with a high elasticity of demand in the variable production coefficient case. See Vernon and Graham (1971), Schmalensee (1973), Warren-Boulton (1974). The innovative semiconductor producer may be seen as a temporary monopolist in certain types of LSI devices. Consumer goods markets may be seen as characterized by many producers and a high elasticity of demand. LSI devices substituted other semiconductor components and mechanical devices previously used in the production of watches and calculators: and this may be seen as a variable coefficient production function.

196. Texas Instruments, for example, left the digital watch business (*The New York Times*, 30 May 1981). Other companies remained in the business, although on a more reduced scale compared to the mid-1970s, because they had acquired marketing skills in the consumer market. For example, a top executive of National Semiconductor claimed: 'National Semiconductor

paid its entrance fee and learned how to market' (*Fortune*, 12 January 1981).

197. For a list, see United Nations (1983).

198. IBM bought 12 per cent of Intel's capital in an attempt to increase its capability in semiconductors. IBM satisfied part of its semiconductor needs in-house and purchased the rest from Intel. IBM's purchases accounted for 13 per cent of Intel's sales in 1981. The decision to strengthen its ties with Intel was taken by IBM because IBM '[realized] that technology [was] accelerating and it [couldn't] keep up if it [did] it all in-house . . . it intends to be aggressive in technology.' (*Business Week*, 10 January 1983.)

199. *Business Week*, 10 January 1983. Intel developed proprietary circuits for Burroughs and sold microprocessors to IBM; Texas Instruments developed devices for communication systems for IBM; National Semiconductor and Motorola developed VLSI devices for a joint venture of fifteen companies.

In addition, two major groups had been established in the computer and semiconductor industry. The Microelectronics and Computer Technology Corporation (MCC) is a non-profit corporation with thirteen members and an annual budget of $75m. The Semiconductor Research Co-operative (SRC) has nineteen members and an annual budget of over $8m. for university research (*Business Week*, 15 August 1983, pp. 94–5).

200. This restructuring caused ATT to seek co-operation in fields which originally were not in its sphere of competence, i.e. consumer electronics and computers. Hence, ATT agreed to co-operate with Philips in 1982 and Olivetti in 1983. In 1983, for example, Western Electric was in the race for introducing the 256K RAM memories, *The New York Times*, 17 July 1983.

201. In the United States, the sale of Japanese private-label products was approximately equal in value to the sale of Japanese brand-label products. In the second half of the 1970s, imports into the United States of TV sets from Japan declined, since beginning in the mid-1970s, Japanese producers had made direct foreign investments in the United States (Peck and Wilson, 1981 and Levy, 1981).

202. After 1973, however, public procurement increased again.

203. In general, during the 1970s technological spillover from commercial R&D to military applications was probably greater than flows from military R&D to commercial applications (see Levin, 1982).

204. In 1979 semiconductor sales accounted for approximately 18 per cent of NEC's total sales, 7 per cent of Fujitsu's, 6 per cent of Toshiba's, 4 per cent of Hitachi's, 4 per cent of Mitsubishi's and 2 per cent of Matsushita's (Borrus, Millstein and Zysman, 1982).

205. Masuda and Steinmuller (1981).

206. The role played by the specific Japanese financial structure will not be examined here. For a discussion, see Flaherty and Itami (1984).

207. Uenohara *et al.* (1982).

208. Nomura (1980).

209. Borrus, Millstein and Zysman (1982).

210. For a discussion of the trade deficit in LSI devices, see Krugman (1983) and (1984).
211. *The New York Times*, 28 February 1982 and 6 September 1984.
212. In the VLSI project (1976–9), for example, NEC, Hitachi and Fujitsu from one side, and Toshiba and Mitsubishi from the other, co-operated in R&D. In the opto-electronics project (1979–86), Fujitsu, Hitachi, NEC, Furukawa and Sumitomo from one side, and Toshiba, Mitsubishi, Matsushita and Oki from the other, co-operated in R&D (Peck, 1983).
213. See Peck (1983). Borrus, Millstein and Zysman (1982) propose the concept of 'controlled' competition in which 'the intensity of competition between firms in key industrial sectors is directed and limited both by state actions and by the formal and informal collaborative efforts of industrial and financial enterprises' (p. 61).
214. Bank of America (1979) and Nomura (1980).
215. Borrus, Millstein and Zysman (1982).
216. Uenohara *et al.* (1982).
217. Nomura (1980).
218. The IBM share of the Japanese computer market has been declining from 37 per cent (1966) to 25 per cent (1976). See Nomura (1980), Bank of America (1979) and *Business Week* (16 July 1984). Japanese firms' share has been much higher than the share held by European computer producers on the European market.
219. For example, in 1973 and 1974 MITI supported R&D in integrated circuits with, respectively, Y1,700m. ($6.1m.) (1973) and Y1,800m. ($6.2m.) (1974). Source: OECD, 1977, p. 203.
220. I thank David Mowery for emphasizing this point for me.

7   Conclusions

1 Supply, demand and public policy factors in the evolution of the European semiconductor industry: a summary

In the previous chapters the evolution of the European semiconductor industry was analysed according to supply, demand and public policy factors. These factors were also used to explain the relative performance of the European industry in comparison with the American and the Japanese industries. The development of the European semiconductor industry is briefly summarized in the following pages.

The initial conditions of the industry in the late 1940s and early 1950s were characterized by a situation of both technological and competitive parity between the European and the American industries, with the Japanese industry behind these two. The capabilities of European firms in electrical and receiving tube technologies were as developed as those of American firms. European firms, moreover, were doing advanced research in solid-state physics and developing certain types of semiconductor diodes, as were American firms.

During the transistor period, supply factors were similar in Europe, the United States, and Japan. Vertically integrated receiving tube producers entered the semiconductor industry because of technological beliefs, economies of scope, and security of supply. These producers eventually became the major producers of semiconductor devices during the 1950s.

During the transistor and early integrated circuit periods, demand and public policy factors had several similarities in Europe and Japan. The structure of demand was dominated by a demand for semiconductors, which was linked with the production of electronics consumer goods. The computer market, on the other hand, was small; in Europe, this market was dominated by American producers. Public policy factors did not play a major role in the development of the domestic semiconductor industry in either Europe or Japan. Specific government policies in support of the domestic semiconductor industry were either limited or totally absent. Unlike in Europe, however, the government in Japan protected the domestic market from the penetration of American producers as the market for integrated circuits grew during the 1960s.

During the transistor and early integrated circuit periods, demand and public policy factors in the United States differed from those in Europe and Japan, and greatly influenced the rate and direction of technological change in the American industry. The structure of demand for semiconductors in the United States had a very large share of public procurement and a growing share of computer demand, while public policy in the United States supported the establishment of both the semiconductor and computer industries. The structure of demand and the policy of the government increased the rate of innovation and channelled technological change in specific directions: the military 'pulled' integrated circuits, while the computer industry 'pulled' digital integrated circuits.

As a result of this combination of demand and policy factors, the structure of the American semiconductor industry was radically altered, and American firms were able to accumulate technological capabilities and productive experience in the new silicon-planar-integrated circuit technological regime during the1960s. New merchant producers who had not been committed to the previous germanium-alloy-mesa-discrete device technology entered the semi-conductor industry and became the major American producers of integrated circuits by the end of the 1960s. These new merchant producers were able to profit from the switch in technological regimes from discrete devices to inte-grated circuits. The routines of these new American merchant producers were quick to adapt to, and innovate in, the new silicon-planar-integrated circuit technology. They were therefore able to increase their market shares within the American industry. American vertically integrated receiving tube producers, on the other hand, changed their routines slowly and were slow to adapt to the new technological regime. In addition, the largest American computer producers integrated vertically upstream into semiconductor production because the size of their internal demand for semiconductors permitted them to do so; most of them remained captive producers. As a result, during the 1960s, a change of firms in the industry (from vertically integrated receiving tube producers to merchant producers and vertically integrated computer producers) also meant a change in routines (from the germanium-alloy-mesa-discrete devices tech-nology to the silicon-planar-integrated circuit technology).

In Europe, on the other hand, the new integrated circuit technological regime was met by the same industry structure that had characterized the transistor period. This structure was composed of vertically integrated receiving tube pro-ducers that were slow to change routines and shift to the new silicon-planar-integrated circuit technology.

As a result, when the computer and digital semiconductor devices markets began to grow in Europe during the 1960s, these markets were penetrated either by imports from the United States or by the newly created subsidiaries of American firms. These American semiconductor firms had technological and productive advantages over existing European producers as a result of the advanced technological capability and productive experience which they had accumulated in silicon-planar-integrated circuit technology. As a consequence, European vertically integrated semiconductor producers were pushed out of the fastest growing digital integrated circuit market and were forced to remain instead in the production of discrete devices and linear integrated circuits for the consumer and industrial markets.

Like European firms, Japanese vertically integrated producers were also late in the shift to the new silicon-planar-integrated circuit technological regime and lagged behind American firms in integrated circuits during the 1960s. However, at this time, there existed a very important difference between the European and the Japanese cases. Unlike the European governments, the Japanese government protected the domestic market from the penetration of American producers.

During the first part of the LSI period (the first half of the 1970s), American firms dominated the world LSI device market. European producers were able to survive in the European semiconductor market by producing discrete devices and linear integrated circuits and by specializing in custom and speciality products. Japanese firms, by contrast, began to accumulate both technological capability and productive experience in MOS LSI integrated circuits as a result of demand (a large calculator demand and a growing computer demand) and public policy factors (the government targeted semiconductors as a strategic industry and favoured R&D co-operation among Japanese firms).

In the course of the LSI period, differences in the structure of demand for semiconductors in Europe, Japan and the United States diminished. The convergence of computers, telecommunications equipment and electronics consumer goods, as well as the digitalization of linear functions, resulted in a rather similar structure of demand for LSI devices in these three major geographic areas by the end of the 1970s.

During the second half of the 1970s and early 1980s, the demand for LSI devices induced a wave of entries by European electronics final goods firms into the R&D and production of LSI devices. These firms vertically integrated in order to remain competitive in their final products—mainly telecommunications, consumer electronics, and industrial equipment. A similar wave of entries occurred in the United States and Japan.

The entry of European firms into the R&D and production of LSI devices was favoured by public policy factors. Unlike previous periods, during the second half of the 1970s and early 1980s, European governments launched several programmes in support of their domestic semiconductor industries. While all of these programmes provided established semiconductor firms with R&D funds and investment subsidies, however, few of them were successful (the exception may be Germany).

In the early 1980s, when the LSI period turned into the VLSI period, European producers that had already entered the LSI market continued to be rather unsuccessful. Despite both the amount of resources which they had committed to the R&D and production of LSI devices and the support which they were given by their domestic governments, these producers were not able to reach the technological frontier or to increase their market shares in the standard LSI device market. Some of these producers, however, were more successful in the more protected and less competitive markets for custom, semicustom, and application specific LSI devices.

The relative lack of success of European firms in the LSI device market was the result both of their late entry and of the character of LSI technology. In fact, because of the complexity and appropriability of LSI technology and because of the cumulativeness of technological advance in LSI technology, a late entry into LSI technology constituted a major disadvantage. Late entrants, such as the European vertically integrated firms, needed time to establish an advanced capability in the new technology and to accumulate productive experience.

The sharing of technological knowledge through R&D co-operation, the continuous building up of an advanced technological capability and the accumulation of productive experience in LSI technology are at the base of the successful performance of Japanese firms in LSI technology in the early 1980s. Because of demand and public policy factors in the second half of the 1960s, Japanese producers were able to commit themselves to LSI technology in the early 1970s. Throughout the course of the 1970s, demand factors (such as an increasingly competitive Japanese computer industry), and public policy factors (such as a co-ordinated, sensitive and protective public policy) continued to support and foster the efforts of Japanese producers in LSI technology. Therefore, by the early 1980s, Japanese, and not European, firms were in a position to challenge the supremacy of American firms in the world market for LSI devices.

2 Supply, demand and public policy factors in high-technology industries

It is now time to go back to Chapter 3 and ask what lessons can be learned from the case of the semiconductor industry. The answers to this question will concern, first, the role of supply, demand and public policy factors in the semiconductor industry. This discussion will develop the findings about the European, American and Japanese industries contained in Chapters 4, 5 and 6. After this discussion, some more general conclusions about the role of supply, demand and public policy factors in high-technology industries will be presented.

2.1 *The case of the semiconductor industry*

The analysis of the semiconductor industry provides a powerful argument for the claim that the effects of supply, demand and public policy factors on the rate and direction of technological change have varied over time and across countries. Changes over time are best exemplified by the American semiconductor industry. As Chapters 4 and 5 have shown, in this industry the effects of demand (i.e. public procurement) and supply (i.e. vertically integrated receiving tube firms) factors on the rate and direction of technological change during the 1950s and early 1960s were quite different from the effects of demand (i.e. computer demand) and supply (i.e. merchant producers) factors during the rest of the 1960s. Differences across countries, on the other hand, are best exemplified by the American and European semiconductor industries during the 1960s. As Chapter 5 showed, in Europe demand factors (the prevalence of consumer demand), supply factors (the prevalence of vertically integrated receiving tubes), and public policy factors (the absence of consistent public support for the industry) had a different effect on the rate and direction of technological change than did demand factors (the prevalence of computer demand), supply factors (the prevalence of merchant producers) and public policy factors (a consistent public support for the industry) in the United States.

The examination of the semiconductor industry also indicates that the specific stage of technological evolution affects the specific role played by demand, supply and public policy. During the evolution of a technology, its attributes (such as characteristics, cost, complexity, appropriability, and cumulativeness) change widely, and thus modify the effects of these three factors. As Chapters 4 and 6 have shown, in the semiconductor industry the effects of the same supply, demand and public policy factors present in the transistor period would not have been the same in the LSI period because technology had changed radically.[1]

The specific stage of technological evolution also affects the ease, speed and mode of both the diffusion and the transfer of technology. This may be best exemplified by a comparison of the transistor and LSI periods. During the transistor period, technology was easy to understand, easy to transmit, and relatively inexpensive to master. In addition, the transfer did not require advanced technological capabilities. The transfer of transistor technology across countries was therefore easy and rapid. The opposite is true for the LSI period. LSI technology is complex to understand, complicated to transmit, and extremely costly to master. In addition, the transferee requires advanced technological capabilities in LSI technology in order to adopt rapidly and successfully the technology transferred. For this reason, the transfer to LSI technology from firms and countries with LSI technological capabilities to firms and countries without LSI technological capabilities has been difficult and slow. In European firms, for example, a long time elapsed between the commitment to LSI technology and the full mastery of this technology.

A comparison of the transistor and LSI periods also exemplifies the change in the mode of technology transfer over time. While licences were the common mechanism of transmission for transistor technology, co-operation, joint ventures and technological exchanges have been increasingly used as transmission mechanisms for LSI technology because of the complexity, cost and appropriability of this technology.

Supply factors: firms structure and strategies

The case of the semiconductor industry has shown how different the structures and strategies of firms have been during the history of the semiconductor industry. As discussed in Chapter 2, these differences are the result of the continuous change, high uncertainty and bounded rationality that have characterized this industry since its beginnings.

The distinction of firms according to age, success and linkages has proved useful only in specific circumstances. As mentioned in Chapter 2, the age of a firm is an indicator of experience and knowledge; it influences the way in which the search process is undertaken. The success of a firm is an indicator of competitiveness in current technologies; it affects the range and intensity of search. The linkages of a firm are indicators of the technological and productive areas of activity of this firm; they affect the focus of search. If we go back to the first 'cut' among new and established firms (see Chapter 3), we can expect that established and successful firms innovate incrementally, while new or unsuccessful firms profit best from radical changes in technology. According to this first 'cut', when technological change is incremental, established and successful firms can best make use of their knowledge, experience and competitiveness in current technologies. By contrast, when technological change is radical, established or successful firms do not have major advantages over new or unsuccessful firms in the innovation process since knowledge, experience and competitiveness in previous technologies is of limited use. In actual fact, by not having any vested capability or interest in previous technology, new or previously unsuccessful firms may grasp and commit themselves to radically new technological opportunities more rapidly than will established or successful firms. This 'cut' proved correct in a specific period during the evolution of the semiconductor industry.

During the integrated circuit period, in fact, new American merchant producers moved quickly and commited themselves forcefully to the new silicon-planar-integrated circuit technology. This rapidity of movement into the new technology was also common to some firms not particularly successful in the previous germanium-alloy-mesa-discrete devices technology. Firms that were established and successful in the previous germanium-alloy-mesa-discrete devices technology (such as American and European vertically integrated receiving tube producers), on the other hand, were reluctant to move either quickly or forcefully into the new technology. In addition, as observed above, in the periods that directly followed the introduction of each of the three radical innovations in the semiconductor industry (the early 1950s, early 1960s, and early 1970s) new firms entered the American industry.

This first 'cut' proved wrong, however, both in the early and in the most recent periods of the semiconductor industry (as well as in other high-technology industries). In the early period of the industry, established and successful firms, such as ATT and GE, introduced several major innovations. In the most recent periods of the history of the semiconductor industry, established and successful firms, such as Intel and Texas Instruments, maintained highly innovative records.

In the semiconductor industry, several factors have enriched and complicated the simple relationship between the age and success of firms and their ability to innovate. First (as mentioned in Chapter 3), in most high-technology sectors established and successful firms are more and more prompt to adapt and commit themselves to new technologies. In addition, interdependencies or relationships between technological regimes, or between major innovations, may allow established and successful firms to make radical innovations.

In the semiconductor industry, for example, the shifts from receiving tubes to transistors and from integrated circuits to LSI devices were accompanied by rather similar technological strategies among established and successful semiconductor producers. In the first shift, most firms understood immediately the importance of the transistor because of the obvious advantages of a solid-state amplifier over the former vacuum tube. In the second shift, established and successful producers were in agreement over the importance of future developments in LSI technology, in part because LSI technology represented the continuation of a trend of increasing integration and miniaturization in digital integrated circuits which had already been established during the 1960s (see Chapter 6). Therefore, as Chapters 4 and 6 showed, in the passage from the receiving tube to the transistor period, and from the integrated circuit to the LSI period, firms that were established and successful in the previous technological regime continued to be major innovators in the next technological regime.

It is interesting to note that in the case of digital integrated circuits most European firms were not only late, but also wrong in their choice of American partners from whom to obtain digital integrated circuit technology. During the early 1960s, in fact, several European firms obtained licences from American vertically integrated receiving tube producers (the losers) rather than from American merchant producers (the winners).

A third factor which has complicated the relationship between the age and success of firms and innovation, is related to situations of low appropriability of innovation and low cumulativeness of technological change which permit established and successful firms to correct mistakes in their choice of technologies easily and quickly, without damaging their competitiveness. In fact, because of the low cumulativeness and appropriability of technological change in the transistor period, even an initial variety in strategies could not harm the performance of firms in the long run: firms that made mistakes could correct them easily, rapidly and inexpensively (see Chapter 4).

Vertical integration

Among the various intersectoral linkages which have affected the semiconductor industry, this case study has highlighted the crucial role played by the vertical integration linkage throughout each period in the history of the industry. This linkage is related to the incentives and consequences of vertical integration at the productive and innovative levels for both input producers and input users.

As discussed in Chapters 4, 5 and 6, the incentives for, and consequences of, vertical integration at the productive level, have been numerous and have varied according to whether vertical integration involves input users or input producers. In the case of input users, the presence of high transaction costs, the possibility of disclosure of proprietary information about electronics final goods producers by semiconductor firms, the growing demand for final products using semiconductors and, in early periods of the industry, the small size of the semiconductor industry, have been strong incentives for upstream vertical integration. In the case of input producers, the high dependence on input, the high value added, and the high elasticity of demand of the final product have been strong incentives for downstream vertical integration. The consequences of vertical integration for input producers and input users, on the other hand, have varied according to the business cycle. During periods of reduced overall demand for input, vertical integration provided a safe outlet for input producers but did not permit input users to profit fully from increased competition on the external market. The situation was reversed, however, during periods of increased overall demand for input. In this case, vertical integration inhibited the in-house input producers from capturing all the new stimuli, opportunities and linkages associated with the growth of demand, while it provided the input users with a guaranteed supply of input.

The history of the semiconductor industry also indicates that the incentives and consequences of vertical integration at the innovative level, unlike at the productive level, are similar for both the input producer and the input user. As mentioned in Chapter 3, these incentives and consequences are related to the increased interaction between, and the improved use of information by, producers and users, as well as to the reduction of user uncertainty for input producers. They become stronger as the product becomes more custom since both producers and users will be able to profit more extensively from their interaction. Input producers will be able to develop inputs which are better suited to the requirements of the users and input users will be able to develop final products which contain highly appropriate inputs. It is important to note, however, that in order to be innovative, both the producers and the users must be at the technological frontier.

As far as vertical integration is concerned, the general proposal advanced earlier has been confirmed: the specific stages of technological development do have varying effects on the types of incentives for, and consequences of, vertical integration. In the transistor period, for example, semiconductor technology remained easy to master, the industry was new, semiconductor producers were few, and electronic final markets were growing rapidly. A fear of supply shortages therefore induced several electronics final goods producers to integrate vertically into semiconductor production. In the LSI period, by contrast, semiconductor technology had increasingly acquired sub-system and system characteristics, the semiconductor industry had become a major industry with a great variety of producers, and electronics final markets had undergone a process of convergence. Because of the high semiconductor content and the high value added of certain types of electronics final goods, some semiconductor producers integrated downstream into the production of these electronics final goods. Moreover, because of the strategic importance of LSI devices for electronics final goods, several electronics final goods producers integrated upstream into semiconductor production.

It should be noted, however, that in the semiconductor industry vertical integration has always represented a very complex linkage. First, because of technical and productive characteristics, in most cases vertical integration has been tapered, both for the input producer and for the input user. As a result, the effects of the external, competitive market could have offset any eventual negative consequences of fully captive in-house production and in-house purchase. Second, in some cases a situation of quasi-integration has developed between producers and users. Quasi-integration allowed firms to avoid the costs of complete vertical integration, without relying exclusively on the market mechanism.[2] Finally, in recent times the vertical integration of the R&D stages, and not the productive stages, has become increasingly common. Therefore, either by the use of silicon foundries or the purchase of semicustom devices, electronics final goods producers have been able to avoid the cost of investments in the full-scale production facilities of semiconductor devices.

Market structure, appropriability and technological change

The case of the semiconductor industry clearly indicates that more than one type of selection process may characterize the evolution of an industry. The selection process may regard firms which embody different routines: firms that embody routines that are most adapted to the specific conditions of the industry will survive. The selection process that occured in the American semiconductor industry is a case in point. Alternatively, the selection process may regard routines within single firms; firms do not change, but their routines do. The selection process that occurred in the European semiconductor industry demonstrates this point.

In addition, the two-way relationship between market structure and technological change has been confirmed by the evolution of the semiconductor industry. As introduced in Chapter 2, in high-technology industries the structure of an industry affects the rate and direction of technological change through the composition of the strategic groups and the conditions of entry into, and exit from, the industry. The rate and direction of technological change in turn affects the structure of the industry, because technological change alters the barriers to entry, the relevance of each strategic group, and the market shares of the various firms in the industry. In the case of the semiconductor industry, during the transistor and integrated circuit periods, the structures of the European and Japanese industries (composed of vertically integrated receiving tube producers with low mortality rates) and of the American industry (composed of a combination of merchant producers and vertically integrated producers with high entry and high mortality rates) had different effects on the rate and direction of technological change in Europe, Japan and the United States (see Chapters 4 and 5). Later in the integrated circuit period and during the LSI period, however, entry rates in the European and Japanese markets began to differ: the European market became increasingly penetrated by American subsidiaries and exports from the United States, while the Japanese market did not. In turn, the change in technology from transistors to integrated circuits reduced the market shares of vertically integrated receiving tube producers in Europe and the United States, and increased the market shares of merchant producers in the United States and American subsidiaries in Europe.[3]

The two-way relationship between market structure and technological change, however, may undergo several modifications during the transition process from an old to a new industry and from one technological regime to another. The case of the semiconductor industry strongly suggests that these transition processes must be carefully analysed. Each transition process, in fact, may have a specificity of its own. While it could be expected that the establishment of a new technological regime or the introduction of some major innovations are associated with radical changes in industry structure and the rise of new firms, this is not always so. In the semiconductor industry, in fact, the initial structure of the industry consisted of a limited number of large established producers (perpetuating the structure of the previous receiving tube industry), and not of many new firms. In addition, the shift from the integrated circuit regime to the LSI regime was characterized by the same group of major firms, and did not provoke a change in identity of the major world semiconductor producers.

Finally, it should be emphasized that the evolution of the semiconductor industry in various countries has been influenced by appropriability conditions. As proposed in Chapter 3, appropriability concerns either the returns on innovation or technological knowledge (that can be used as an input in the innovative process). The appropriabilty of the returns on innovation remained high in the United States, Japan and Europe. The appropriability of technological knowledge, by contrast, differed among these countries. In the United

States, the high mobility of technicians and R&D personnel among firms reduced the ability of single firms to appropriate technological knowledge. A similar situation occurred in Japan as the result of the co-operation of firms in R&D. In both countries, the low appropriability of technological knowledge meant that domestic firms enjoyed a diffused and rather even level of applied technological knowledge, which they were able to exploit in later innovations. Nothing of this sort occurred in Europe, where the appropriability of techno-logical knowledge by single firms remained high.

Intersectoral linkages and demand factors

One of the major conclusions that may be drawn from the previous chapters concerns the importance of intersectoral linkages in shaping the evolution of an input industry such as the semiconductor industry. These linkages, in fact, have worked to connect semiconductor firms with downstream industries with different characteristics. Firms that were linked with the computer industry were subject to innovative and productive stimuli that were more dynamic than those to which firms linked with the electronics consumer goods industry were subject.

Among the various demand factors, the structure of demand has played a central role in the semiconductor industry. Changes over time and differences across countries in these linkages, and therefore in the structure of demand, have greatly influenced the rate and direction of technological change in the semiconductor industry. This influence is most obvious in the case of the diversity of the demand structure between Europe, Japan and the United States during the 1960s (military and computer vs. consumer goods demand), as Chapter 5 has shown. This influence was confirmed during the 1970s, when the convergence of electronics final markets took place (see Chapter 6).

This convergence affected the various electronics final goods producers in different ways. IBM, for example, had to create new horizontal intersectoral linkages with sectors such as telecommunications equipment and factory auto-mation, and to strengthen its vertical intersectoral linkages in LSI semiconductor technology.[4] Similarly, ATT had to create new horizontal intersectoral linkages with consumer products, office equipment and computer sectors.[5] Among semi-conductor producers, most European producers had to modify their previous linkages with either the consumer or the telecommunications equipment market.[6]

Public policy factors

The case of the semiconductor industry demonstrates the influence of pub-lic policy on the rate and direction of technological change as well as on the international competitiveness of firms and countries. Public policy acted as a positive force in the United States and Japan and as a neutral force in most European countries. American government policy toward the semiconductor industry was successful early in the history of the industry. During the 1950s and early 1960s, it increased the rate of technological change, affected the direction of innovation toward silicon integrated circuits, and contributed to the establishment of the technological and commercial superiority of the American industry. The success of government policy, however, was favoured by two unique conditions: the semiconductor industry was in its infancy and the requirements of military demand coincided with the requirements of civilian demand. Japanese government policy was successful later in the history of the industry. During the 1970s, it helped Japanese producers to close the gap with American producers in several semiconductor products and technologies. European government policies, on the other hand, were unsuccessful in their aim to close the gap with American producers.

However, the history of the semiconductor industry also indicates how the policies of the various countries have varied in terms of goals, tools, locus of decision, timing and flexibility. American military policy toward the semiconductor industry established clear technological targets for semiconductor devices with well-defined characteristics, remained flexible and sensitive, promoted competition among firms with alternative projects, and favoured the entry of new firms into the industry. Japanese policy toward the semiconductor industry targeted specific areas of support and remained flexible and sensitive; contrary to the American policy, however, it organized and co-ordinated the R&D activity of established Japanese firms, thus avoiding duplicative R&D efforts. In addition, both American and Japanese policies protected domestic markets: American military policy, through various contract clauses and national security measures, and Japanese policy through tariff and non-tariff barriers. Finally, while unable to protect the domestic market wholly, German policy possessed some characteristics similar to the American and Japanese cases; in particular, it established clear technological targets and remained attentive to the technological strategies and capabilities of established German semiconductor producers. By contrast, the policies of France, and to a lesser extent Great Britain, during the 1970s, were closer to the second type. The French government, in fact, often intervened directly in the strategies of French firms and in the structure of the domestic semiconductor industry. While less interventionist than French policy, British policy also intervened directly in the structure of the domestic industry on occasion. In addition, most French and British policies were spread over a wide range of products and did not interrelate with policies in other sectors.

2.2 *Generalization for high-technology industries*

To what extent can the preceding conclusions about the semiconductor industry be generalized for all high-technology industries? Some of them can indeed be framed into a broader perspective and these will be discussed briefly below. Others, however, remain sector-specific, and must be considered only in this perspective. This aspect will also be discussed briefly in the following pages.

The findings of this study on the semiconductor industry add up to a general perspective on high-technology industries. This perspective considers technological features, as well as supply, demand and public policy conditions as major factors in the determination of the rate and direction of technological change over time and across countries. The perspective can be associated with that part of the literature on technological change which considers both 'technology push' and 'demand pull' as important factors in affecting the rate and direction of technological change.

In addition, because changes in technology include both innovation and diffusion processes, it follows that both these features and these factors must be taken into consideration in analyses of diffusion processes and of the transfer of technology over time and across countries. Some of these features and factors have been emphasized in the general literature concerning the diffusion process[7] and the transfer of technology.[8] According to this literature, in fact, technological features (the characteristics, cost and rate of change), appropriability conditions, supply factors (expectations, previous and competing technologies, and industry structure), demand factors (the size and growth of demand) and public policy factors (the size, type and timing of public policy) greatly affect the speed and mode of the diffusion process.

It is suggested here that the technological choices of firms also play a major role in affecting both the rate and direction of technological change and competition among firms and among countries in high-technology industries. As discussed in Chapter 2, these choices differ across firms and are related to the subjective and limited rationality of firms concerning present and future changes in technology. They determine many of the strategies of firms and shape many of their routines. Where both the cumulativeness and the appropriability of technological change exist, firms with the right technological choices are more likely to grow rapidly and to increase their market shares. As a result of learning curves and the appropriation of innovation, in fact, those firms that are first to move in the 'right' direction and the quickest to imitate the 'right' technology will be able to establish and maintain technological and productive advantages over their competitors.

When shifts in technological regimes are radical, technological beliefs and choices may vary widely. As a result, choices have a wide margin for error; the uncertainty surrounding alternative technological options is very high and experience from involvement in previous technologies may be of limited use to firms. Even when shifts in technological regimes are radical, however, the diversity of technological beliefs is low if the continuities and inter-dependencies between the regimes are high and if the new technological regime is soon considered to be superior, both technologically and economically.

Another major result obtained from the former analysis is that, in an input industry, the type of intersectoral linkages present affect the type of techno-logical change that may occur. As mentioned in Chapter 3, intersectoral linkages may be internal (vertical integration) or external (market) to the firm; they may be public (public procurement) or private (civilian market); and they may con-nect the high-technology industry with quite different sectors.

The first type of intersectoral linkage—the degree of vertical integration of firms—affects the rate of technological change according to the features of technology, the dynamics of the industry, and the characteristics of customers and suppliers. These factors, in fact, influence the incentives for, and the direc-tion and the degree of, vertical integration. As a result, the structure of the industry, as well as the consequences of vertical integration at the innovative and diffusion levels, will also be affected. The change in the incentives for and consequences of vertical integration over time, demonstrated here for the semiconductor industry, have been confirmed for a series of industries at dif-ferent stages of development and with different structural characteristics.[9] From this evidence, it is possible to argue that vertical integration should be analysed not only in a static way according to criteria based on efficiency and short-term profitability, but also in a dynamic fashion according to criteria based on industry and technology development,[10] and on the innovativeness of firms.[11]

Equally important are the effects of the specific combination of sectors that are connected by intersectoral linkages on the rate of technological change. In the case of final goods industries, upstream sectors can have a wide range of effects on downstream sectors, according to the importance of upstream pro-ducts in the final goods. In the case of input industries (as shown previously for the semiconductor industry), downstream sectors may or may not be driving forces, depending on the size, growth and type of customers and on the dyna-mics of technology. Firms that are linked with driving sectors will experience more rapid technological change than firms that are linked with non-driving sectors. Owing to their size and dynamics, in fact, driving sectors will pull and stimulate innovation at the input level more than will non-driving sectors.

It is important to note once more that the evolution of the various user industries may affect the rate and direction of technological change in the input or the capital goods industry through a wide range of channels. These channels include not only the size of the input–output linkages among sectors, but also more subtle pulls such as the subjective expectations of producers about the future evolution of user industries and the interaction between the producer and user industries.

It is also possible to conclude that, in an input industry, changes in demand structure over time and diversities in demand structure across countries affect the rate and direction of technological change. These changes and diversities reflect the evolution of the various user industries over time and across countries. They influence technological change because they require products of different types, which use different technologies and which have different characteristics.[12] Firms will therefore be subject to different innovative pulls and loci,[13] deriving from the change or diversity in the structure of demand, and will focus their innovative efforts in different directions over time and across countries.

It must be noted, however, that in industries with a high rate of technological change, the effects of changes over time and diversities across countries in the structure of demand should be distinguished from the effects of the size and growth of overall demand. While changes in the size of demand affect the rate, but not the direction of technological change, changes in the structure of demand affect both the rate and direction. The growth in demand is related to the 'demand pull' approach, as Mowery and Rosenberg (1979) have pointed out. Given a certain size, the growth in demand will affect the innovative effort and the rate of innovation of firms by influencing the expected profitability of successful innovations. When this growth is expected to take place at some future date, the various subjective expectations of firms play an important role in affecting the rate of technological change, as Rosenberg (1982) has noted for the diffusion rate of new technologies. Changes and diversities in the structure of demand, on the other hand, point to a composition effect that can affect not only the rate, but also the direction of technological change. Given the growth rates of the various intersectoral linkages, different compositions of the structure of demand will pull innovation in different directions or will have different effects on the rate at which technologies are diffused.

It should be noted that changes over time in the composition of demand may involve major variations in the structure of the input industry. This is a consequence of the shift of input producers from declining to growing demand segments. Input producers connected with declining segments may in fact lose market shares to producers connected with growing segments during the adaptation of their former capability to the new intersectoral linkage—a process which may require time and resources.[14]

Changes in demand structure across countries may also help to differentiate domestic firms from foreign subsidiaries in the innovation process. This aspect has already been discussed in Chapter 3 for a high-technology industry, and does not need to be repeated here. It must be noted, however, that demand structure will have a different effect on the direction of technological change in domestic firms and foreign subsidiaries only if the initial conditions in the two countries differ significantly and if development follows two different paths.

Another major conclusion which can be drawn from this study on the semiconductor industry, as well as from other studies regarding other strategic or high-technology industries,[15] concerns the distinction between two broad policy types. One type considers firms to be the main actors in the innovation process and the main locus of technological knowledge, regards R&D activity as an uncertain process, and is concerned about the possible duplicative efforts of firms. This policy type has clear technological objectives but does not specify them in detail; rather, it pursues them with flexibility and sensitivity. It attempts to select or co-ordinate the innovative activity of firms or to foster co-operation in R&D. It relies on the market mechanism and lets firms choose the best ways to reach established objectives.

According to the second type of policy, the government is a major actor in the innovation process, and the R&D activity of firms is a process which can be guided and predicted in advance with a certain amount of accuracy. This policy establishes clear technological objectives, specifies them in detail and indicates ways by which to reach them. It relies only partially on the market mechanism, intervening directly in the strategic choices of firms and the structure of the industry.

It is important to note that these two types of policies, while explaining some of the characteristics of the policies of each country, do not fit any of them exactly. It is possible to claim that, at a very general level, American policy toward high-technology industries is closer to the first type, while French public policy is closer to the second type. German and British policies lie somewhere between the first and the second type of policies, the German being closer to the first type and the British being closer to the second.

More generally, it should be noted that an accurate distinction between results which are sector-specific and results which are more general should always guide the analysis of a single high-technology industry. One common mistake, in fact, is to apply indiscriminately to all high-technology industries the results obtained from a single industry study. This error derives from the fact that while both the Schumpeterian process of 'creative destruction' and high rates of change are common to all high-technology industries, in many other respects these industries differ quite extensively. As Nelson and Winter point out, 'firms are not all alike. . . . The extent of the rewards and penalties, and the rates of introduction and diffusion of new techniques, depend on a complex of environmental and institutional considerations that differ sharply from sector to sector, country to country and period to period' (Nelson and Winter, 1974, p. 903). In the study of technological change, therefore, it is necessary to 'move from highly aggregated to highly disaggregated modes of thinking . . . from the general to the specific, from "Technology" to "technologies" ' (Rosenberg, 1976, p. 2).

It is sufficient to compare the characteristics of semiconductor technology with those of the pharmaceutical,[16] aircraft[17] or nuclear[18] technologies, in order to appreciate these differences. Similarly, it is sufficient to review the discussion about the importance of technology push and demand pull factors in several high-technology industries (Freeman 1982) in order to obtain a general idea of the differences in the determinants of innovation across sectors.

Differences across high-technology industries do not only refer to the specific characteristics of their technologies, however. They also refer to the level of appropriability and cumulativeness of technological change. Appropriability conditions at the patent level in the semiconductor industry, in fact, were much lower than those in other high-technology industries: patents were unable to protect innovation in semiconductor technology, while in the chemical and pharmaceutical industries they have in fact been able to do so.[19] Similarly, while technological change in the semiconductor and aircraft industries has been highly cumulative, in other high-technology industries, such as pharmaceuticals, it has not been so.[20] These differences in appropriability and cumulativeness in turn result in very different evolutions in the industries concerned.[21]

Caution should also be exercised when extending the mechanisms related to supply and demand factors in an input or a capital goods industry to high-technology consumer goods industries. These industries, in fact, present quite different characteristics in terms of forward linkages, since their demand is composed of consumers, not firms. Differences in demand within and across countries will, therefore, be related to diversities in habit, culture, income, and sophistication, at the consumer level.[22]

Diversities among technologies and sectors also delimit the types of general conclusions that can be made about the tools and the results of public policy in high-technology industries. The characteristics of successful public policies have varied not only over time and across countries, as in the case of the semiconductor industry, but also from one high-technology industry to another. Differences in the success of various types of public policy have been confirmed by a wide range of studies.[23]

What about firm-level analyses? Here, also, the specific characteristics of various firms, and not the general characteristics of a single representative firm, go furthest to explain the dynamics of high-technology industries in most cases. The various cases discussed in Freeman (1982) concerning single firms in the synthetic materials and electronics industries illustrate this point. Again, the importance of the firm as the major unit of analysis, and of history as the major source of analysis, are brought to the fore.[24]

The overall conclusion that is to be drawn from the former discussion points to a well-defined direction for analysis, as discussed in Chapter 1. The analysis of high-technology industries and the discussion of sector-specific results should not be placed within a static framework. Rather, it should include a dynamic, historic and evolutionary perspective that is able to grasp all aspects of the diversities and changes, continuities and discontinuities that play such a major role in all high-technology industries.

3 A look at the future: the VLSI period

With the 1980s, a new period in the history of the semiconductor industry has begun: the Very Large-Scale Integration (VLSI) period. This period was initiated as a result of technological advances that allowed over 100,000 circuit elements to be crammed onto a single semiconductor device. This period is characterized by semiconductor devices with sub-micron channel width, by microprocessors of 32-bit units or more, and by dynamic RAM memories of 256K of information or more.[25]

As far as technology is concerned, the VLSI period is more a continuation of the LSI period than a rupture with it. Unlike the transistor and the integrated circuit periods, which established radically different technological regimes, the VLSI period differs from the LSI period only in terms of the degree of integration obtained by semiconductor devices. The technological trend of the VLSI period, in fact, is toward the increasing miniaturization of already miniaturized semiconductor devices and towards the increasing integration of already highly integrated functions performed by semiconductor devices. This trend has accentuated the system character of semiconductor devices, the technological interdependencies between these devices and computers, and the convergence among computers, electronic consumer goods and telecommunications equipment. Therefore, most of the supply and demand factors discussed previously for the LSI period will continue to exert effects during the VLSI period.

In order to be successful, however, firms will not only have to invest considerable resources; they will also need time to build up technological capabilities. In fact, the success of firms in VLSI devices will be determined not only by their present-day commitments and their most recent choices of technologies, products and processes, but also by their technological and productive capabilities, some of which have been accumulated in the past. The characteristics of VLSI technology, moreover, will considerably increase the cost of doing R&D at the technological frontier. As a consequence, semiconductor firms will continue to establish R&D co-operation agreements with other semiconductor firms that are specialized in complementary technological areas in order to show new technological knowledge. At the demand level, because of the convergence of computers, telecommunication equipment and electronics consumer goods, firms will face the same demand structure in Europe, the United States and Japan. This homogeneity of demand will increase the process of global internationalization of both firms and markets. Major firms will increasingly produce for a world demand.[26]

At the same time, because of the crucial importance of VLSI devices for the competitiveness of firms at the final goods level, several electronics final goods producers will continue to be vertically integrated into, and commit considerable resources to, the R&D and production of VLSI devices. Because of the complexity, cumulativeness, and appropriability of VLSI technology, however, those electronics final goods producers that will eventually vertically integrate, will do so more through the acquisition of existing semiconductor producers than through internal growth. Because of the cost of investments in production facilities for VLSI technology, moreover, some electronics final goods producers will make use of silicon foundries and semicustom devices. They will not establish a production capability in VLSI technology, but will maintain only the R&D stage.

As far as Europe is concerned, European firms will try to survive in the world VLSI devices market. They will face competition not only from American multinational firms, however, but also from highly competitive Japanese firms. Presently, most European producers are not producing for the standard VLSI market. The majority of European firms continue to focus on custom, semicustom or application-specific VLSI devices; few, mainly Siemens and INMOS, presently target the standard VLSI devices market.

Given the cost and complexity of semiconductor technology, European firms will attempt to increase their R&D co-operation with other European American and Japanese firms. Examples of such agreements can be found in the new arrangement established between Philips and Siemens and between SGS-ATES and Toshiba. These agreements aim to reduce the cost of obtaining technological knowledge, while preserving high appropriability on the returns of innovation. Given the low degree of labour mobility in Europe compared to the United States, these agreements represent a way to diffuse technological knowledge among firms, and to follow the successful experience of the Japanese producers. It should be noted, however, that several American producers as well have set up R&D co-operatives for the same reasons.

As during the LSI period, most European producers are vertically integrated firms. The in-house R&D and production of VLSI devices by these firms is based on technological beliefs concerning future developments in VLSI technology and on the increasing strategic importance of an in-house capability in VLSI technology. VLSI technology is seen as a key component in the information technology in which European firms want to compete. Information technology is considered so important by these European firms, in fact, that they have recently entered new markets, such as office equipment, factory automation and private telecommunications networks.

Even if the performance of most European producers has recently improved, it is difficult to predict their long-term future performance. One of the reasons for their recent success, in fact, is related to the high rates of semiconductor demand following the 1981–2 recession. The high growth rate of semiconductor demand has meant that possible outlets for the semiconductor production of European firms has expanded and that the degree of rivalry between existing competitors has been reduced.

In addition, forecasts about the long-run performance of European firms in the standard VLSI market should be kept separate from predictions about their performance in the custom, semi-custom and application-specific VLSI markets. Predictions about the performance of European producers such as Siemens and INMOS in the standard memory market are difficult to make. A successful performance in the standard VLSI market requires continuous commitment, productive experience and advanced technological capability. While INMOS and Siemens have recently demonstrated a commitment to VLSI technology, they are still in the process of building up productive experience and advanced technological capability. This process will require time. Predictions about the performance of European producers in the custom, semi-custom and application-specific semiconductor devices are easier to make. Given a continuous commitment to VLSI technology, these producers will be successful if they are able to exploit fully their vertically integrated structures and the advantages of better communications and interaction between in-house producers and in-house users. Success, however, requires that these producers remain competitive at the final product level. Evaluations and predictions about the future performance of European firms in the VLSI market, however, should be kept dynamic and relative. They should be dynamic because the frontier in VLSI technology is continuously changing and firms are focusing on technological targets which move with time. They should be relative because the performance of European firms cannot be evaluated in absolute terms, but must be compared to the performance of their American and Japanese competitors.

A more general prediction about the future performance of European firms can be made on the basis of a specific linkage and a specific demand factor. The performance of European semiconductor producers will be improved if a competitive European computer industry develops. In this case, an additional large, growing, and innovative demand for LSI devices will be able to pull the development and the production of VLSI devices in Europe.

In the coming years, public policy factors may also play an important role in improving the performance of the European VLSI semiconductor industry. As previously mentioned, however, European public policies should not replicate previously successful American or Japanese policies, since present developments in semiconductor technology and the competitive position of the European industry are quite different from past conditions.

In 1983 and 1984, new and wide-ranging programmes of support for the industry have been launched by various European governments. The German, French and British programmes have been examined in the previous chapter.[27] These European government programmes represent major improvements over previous programmes, not only in terms of size, but also in terms of content. They support basic research and generic technologies so that several domestic semiconductor producers can benefit from an established advanced technological base for their more proprietary, firm-specific R&D. These programmes also encourage co-operation among domestic firms, so that the cost of R&D is reduced and the duplication of effort avoided. In addition, they target specific areas of firm R&D for support, so that government support is not spread too widely across a broad range of technologies and projects. Finally, these programmes support R&D and production in VLSI devices within a broader framework centred on information and electronics technologies, so that the linkages and interdependencies between VLSI technology and the rest of the electronics industry can be fully exploited. These government programmes point in the right direction and have the possibility of improving the technological capabilities of the European industry.

In addition, in early 1984, the EEC launched a ten-year programme entitled ESPRIT.[28] This programme aims to improve the technological base and the competitive performance of European producers in five major areas: VLSI technology, software, information technology, office equipment and CAD/CAM. The programme plans to devote UC 1,500m. (approximately $1,290m.) during the period 1984-8 in support of selected co-operative R&D ventures among the firms of various member countries. In particular, within VLSI technology this programme supports co-operative R&D in sub-micron MOS and bipolar integrated circuits, CAD, compound semiconductors, and optoelectronics devices.

The ESPRIT programme focuses on the diffusion of technological knowledge among firms, while preserving the appropriability of the returns of innovation at the firm level. In this sense, it tries to copy the successful Japanese experience of firm co-operation in R&D. In addition, it attempts to create a 'European' semiconductor industry, through incentives for co-operation and through the co-operation of the R&D activites of firms of different nationalities. The task of ESPRIT is not an easy one. Each European government has some 'national champion' which it is determined to protect and support. In addition, tensions may arise between the ESPRIT programme and the independent programmes of the various European countries examined above which seek to support their domestic semiconductor industries.

In conclusion, it is possible to argue that at a general level public policies should be flexible and well timed. In industries with a high rate of technological change, such as the semiconductor industry, industry structure and market conditions are in a perennial state of change and technological targets are continuously moving and evolving. As a result, the technological and economic targets of government policy cannot always be defined with accuracy. Instead, those who draft policies should have a clear idea of the direction in which they intend to push both technologies and industries, and should stress adaptation and response to the changes and developments that take place during the implementation of these policies.

Most importantly, government policies should have a historical and comparative vision that enables them to learn from past experiences and to adapt other countries' lessons to the present technological and structural conditions of the industries targeted for support. This historical and comparative vision is necessary because technologies and industries evolve over time, and are shaped by the specific economic and institutional conditions of each country. However, while public policies should learn from the past, they should not simply replicate past experiences. If history rarely offers easy answers, it surely never repeats itself.

Notes

1. It is worth noting that when the focus of analysis shifts from the industry level to product and firm levels, the dynamics of a technology also interact closely with the specific dynamics of products and firms.
2. For a general discussion, see Porter (1980) and Harrigan (1984).
3. It must be noted that these effects of technological change on industrial structure also hold at the product level. LSI technology, for example, affected the structure of the American and European industries by inducing American LSI merchant producers to integrate downstream into the production of electronics consumer goods, and American and European electronics final goods producers to integrate upstream into the production of LSI devices.
4. In creating and fostering these linkages, IBM followed a strategy of acquisition or participation in established successful producers such as Rolm (telecommunications equipment) and Intel (LSI semiconductor producer).
5. In creating these linkages, ATT followed a strategy of agreements with Philips (consumer products and office equipment) and Olivetti (office equipment and computers).
6. Some of these European producers, successful in the production of linear integrated circuits within these previous linkages, were less rapid in the change to the new linkage and to the consequent production of LSI devices.
7. See particularly Mansfield (1961), Rosenberg (1976), Davies (1979), Gold (1981), Stoneman (1983) and Metcalfe (1983).
8. In particular by Teece (1977), Mansfield (1982), Rosenberg (1982) and Pavitt (1984).
9. See, for example, the various cases in Harrigan (1984) and the case of the aluminium industry in Stuckey (1983).
10. In the spirit of the seminal contributions of Stigler (1951) and Adelman (1955).
11. Linking vertical integration to innovation according to the concept of strategic groups (Porter, 1980 and 1981) and to the consequences of producer–user interaction (Von Hippel, 1978 and 1982).
12. At a general level, the discussion about product characteristics can be related to Lancaster (1979), Teubal (1979) and Saviotti and Metcalfe (1984).
13. The basic reference is Von Hippel (1982).
14. The transition from a declining to a growing linkage may be hampered by the specific competitive position of firms in the declining linkages. Unless they have routinized the scanning for new opportunities and new markets,

established and successful input producers which are linked with a declining industry will tend to remain connected to that linkage longer than will unsuccessful firms.

15. Nelson (1982, 1984), Zysman (1977), Borrus, Millstein and Zysman (1982), Krugman (1984).
16. See, for example, Schwartzman (1976) and Grabowski and Vernon (1982).
17. See, for example, Phillips (1971) and Mowery and Rosenberg (1982).
18. See, for example, Keck (1981).
19. See Taylor and Silberston (1973), Mansfield, Schwartz and Wagner (1981) and Levin, Klevorick, Nelson and Winter (1984).
20. For a detailed discussion of cumulativeness in non-high-technology industries, see Sahal (1981).
21. As it has been explored at the theoretical level and with different approaches by Loury (1979), Dasgupta and Stiglitz (1980a and 1980b), Reinganum (1982) and Nelson and Winter (1982).
22. The consequences for the dynamics of competition in this type of high-technology industry are best exemplified in the electronics consumer goods industry. For this industry, see CRA (1979) and Peck and Wilson (1982).
23. See, for example, Keck (1981), Nelson (1982 and 1984) and Zysman and Tyson (1983).
24. In the spirit of Cyert and March (1963), Simon (1979 and 1984), Usher (1954) and Rosenberg (1976 and 1982).
25. For a general discussion, see Rosenberg and Steinmueller (1982).
26. Recently, marketing functions have played an increasingly important role in the competitive race among semiconductor producers. This aspect will not be discussed here. On the subject, see Flaherty (1983 and 1984).
27. See Chapter 6, Section 5. In addition, Sweden has also recently launched a major programme in support of the microelectronics industry.
28. European Strategic Program on Research in Information Technology.

Bibliography

Abernathy, W. J., *The Productivity Dilemma: Roadblock to Innovation in the Automobile Industry*, Baltimore, Johns Hopkins University Press, 1978.

Abernathy, W. and J. M. Utterback, 'Patterns of industrial innovation', *Technology Review*, **80**, June–July 1978, pp.. 2–29.

Adams, P., 'Government–Industry Relations in Italy: the Case of Industrial Policy', doctoral dissertation, Yale University, 1985.

Adelman, M., 'Concept and Measurement of Vertical Integration', in *Business Concentration and Price Policy*, Princeton, Princeton University Press, 1955.

Alchian, A. A., 'Uncertainty, Evolution and Economic Theory', *Journal of Political Economy*, **58** (1950), pp. 211–22.

— and H. Demsetz, 'Production, Information Costs and Economic Organization', *American Economic Review*, **62** (1972), pp. 777–95.

Allison, D., (ed.), *The R&D Game: Technical Men, Technical People and Research Productivity*, Cambridge, MA, MIT Press, 1969.

Amato, N., *Ricerca Scientifica e Tecnologica: Rapporto del Ministero del Bilancio su 'Attività concernente il governo dell'industria'* (a cura di Riccardo Gallo), Rome, 1983.

Antonelli, C., and B. Lamborghini, *Impresa pubblica e tecnologie avanzate*, Bologna, il Mulino, 1978.

Arnold, E., 'Competition and Technological Change in the U.K. Television Industry', doctoral dissertation, SPRU, Unversity of Sussex, 1983.

Arrow, K., 'Economic Welfare and the Allocation of Resources for Inventions', in Nelson R. (ed.), *The Rate and Direction of Inventive Activity*, Princeton University Press, Princeton, 1962.

Arrow, K. J., *The Limits of Organization*, New York, Norton, 1974.

Arrow, K., 'Vertical Integration and Communication', *Bell Journal of Economics*, Spring 1975.

Asher, N. J. and L. D. Strom, 'The Role of the Department of Defense in the Development of Integrated Circuits', IDA Paper P-1271, Arlington, VA, Institute for Defense Analysis, May 1977.

Atkinson, A. and J. Stiglitz, 'A New View of Technical Change', *Economic Journal*, **79** (1969), pp. 573–8.

Bakis, H., *IBM: une multinationale régionale*, Grenoble, Université de Grenoble, 1977.

Baloff, N., 'The Learning Curve—Some Controversial Issues', *Journal of Industrial Economics*, **14** (July 1966), pp. 275–82.

Bank of America, *The Japanese Semiconductor Industry: an Overview*, 1979.

Baumol, W. J., J. C. Panzar and R. D. Willig, *Contestable Markets and the Theory of Industry Structure*, New York, Harcourt Brace Jovanovich, 1982.

Behrman, J. N., *Some Patterns in the Rise of the Multinational Enterprise*, Graduate School of Business, University of North Carolina, 1969.

Borrus, M., J. Millstein and J. Zysman, *U.S.-Japanese Competition in the*

Semiconductor Industry, Berkeley, CA, Institute of International Studies, University of California, 1982.

Braun, E. and S. MacDonald, *Revolution in Miniature*, Cambridge, Cambridge University Press, 1982.

Breitenacher, M., K. D. Knordel, D. Schedl and L. Scholz, *Elektrotechnische Industrie*, IFO-Institute für Wirtschaftsforschung, Munchen, Duncker-Humblot, 1974.

Brezzi, P., *La Politica dell'Elettronica*, Rome, Editori Riuniti, 1980.

Brock, G., *The US Computer Industry*, Cambridge, MA, Ballinger, 1975.

Burns, T. and G. Stalker, *The Management of Innovation*, Tavistock, London, 1961.

Carter, C. (ed.), *Industrial Policy and Innovation*, London, Heinemann, 1981.

Caves, R., 'International Corporations: The Industrial Economics of Foreign Investment', *Economica*, **38** (February 1971), pp. 1–27.

— 'Corporate Strategy and Structure', *Journal of Economic Literature*, **18** (1980), pp. 64–92.

—, *Multinational Enterprise and Economic Analysis*, Cambridge, Cambridge University Press, 1982.

— and M. E. Porter, 'From Entry Barriers to Mobility Barriers', *Quarterly Journal of Economics*, **91** (1977), pp. 241–61.

Censis, 'L'industria elettronica in Italia', *Censis,* **52, 53, 54**, Rome, 1967.

Chandler, A., *Strategy and Structure*, Cambridge, MA, MIT Press, 1962.

—, *The Visible Hand: The Managerial Revolution in American Business*, Cambridge, MA, Harvard University Press, 1977.

Chang, Y. S., 'The Transfer of Technology: Economics of Offshore Assembly: The Case of the Semiconductor Industry', College of Business Adminstration, Boston University, UNITAR, 1971.

Charles River Associates, *International Technological Competitiveness: Television Receivers and Semiconductors*, CRA, Boston, MA, 1979.

Coase, R., 'The Nature of the Firm', *Econometrica*, **4** (1937), pp. 386–405.

Cyert, R. M. and J. C. March, *A Behavioral Theory of the Firm*, Englewood Cliffs, NJ, Prentice Hall, 1963.

Dasgupta, P. and J. Stiglitz, 'Industrial Structure and the Nature of Innovative Activity', *Economic Journal*, **90** (1980a), 266–93.

—, 'Uncertainty, Industrial Structure and the Speed of R–D', *Bell Journal of Economics*, **11** (1980b), pp. 1–28.

Davies, S., *The Diffusion of Process Innovations*, Cambridge, Cambridge University Press, 1979.

Day, R. H. and T. Groves, *Adaptive Economic Models*, New York, Academic Press, 1975.

Dickson, K., 'The Influence of Ministry of Defense Funding on Semiconductor Research and Development in the United Kingdom', *Research Policy,* **12**, No. 2 (1973), pp. 113–20.

Diodati, J., 'Vertical Integration as a Determinant of Industry Performance in the Telecommunications Industry: An International Comparison', paper presented at the EARIE Conference, Bocconi University, Milan, September 1980.

Dosi, Giovanni, 'Technical Change, Industrial Transformation and Public Policies:

The Case of the Semiconductor Industry', Sussex European Research Centre, University of Sussex, 1981.
— 'Technological Paradigms and Technological Trajectories—A Suggested Interpretation of the Determinants and Directions of Technical Change', *Research Policy*, 11 No. 3 (1982), pp. 147–62.
Dummer, G. W. A., *Electronic Inventions from 1745 to 1976*, New York, Pergamon Press, 1978.
Dunning, J., 'The Determinants of International Production', *Oxford Economic Papers*, 25, No. 3 (1973), pp. 289–336.
— 'Trade, Location of Economic Activity and the Multinational Enterprise: A Search for an Electric Approach' in B. Ohlin, P. O. Esselborn and P. M. Wijkinan (eds), *The International Allocation of Economic Activity*, New York, Holmes-Meier, 1977.
— 'Explaining Changing Patterns of International Production: In Defense of the Eclectic Theory', *Oxford Bulletin of Economics and Statistics*, 41 (1979), pp. 289–96.
Electronic Industries Association, *Electronic Market Data Book*, Washington, Electronic Industries Association, various years.
Enos, J. L., 'Invention and Innovation in the Petroleum Refining Industry', in *The Rate and Direction of Inventive Activity-Economic and Social Factors*, US Bureau of Economic Research, Princeton, Princeton University Press, 1962, pp. 549–83.
Ergas, H., 'Public Utilities and Industrial Policy: The Case of French Telecommunications Policy', paper presented at the seminar on 'Industrial Policy and Structural Adjustment', Naples, April 1983.
Ernst, D., *Restructuring World Industry in a Period of Crisis—The Role of Innovation*, UNIDO Working Paper on Structural Change, Vienna, UNIDO, 1981.
Fast, *Rapporto sulla microelettronica nazionale*, Milano, Fast, 1980.
Finan, W., *International Transfer of Semiconductor Technology through US-Based Firms*, New York, National Bureau of Economic Research, 1975.
— 'The Semiconductor Industry's Record on Productivity', in *American Prosperity and Productivity: Three Essays on the Semiconductor Industry*, Cupertino, CA, Semiconductor Industry Association, 1981.
Federal Republic of Germany, Bundesministerium für Forschung und Technologie (BMFT), *Program Elektronische Bauelemente 1974-78*, Bonn, 1974.
— *6th Report of the Federal Government on Research*, Bonn, 1980.
— *Leistungsplan O4*, Mikroelektronik 1981, Bonn, 1981.
— *Informationstechnik*, Bonn, 1984.
Flaherty, M. T., 'Prices versus Quantities and Vertical Financial Integration', *Bell Journal of Economics*, 12, No. 2 (1981), pp. 507–25.
— 'Market Shares Determination in International Semiconductor Markets', Graduate School of Business Administration, Harvard University, 1984, No. 4.
— 'Field Research on the Link between Technological Innovation and Growth: Evidence from the International Semiconductor Industry', Graduate School of Business Administration, Harvard University, 43, 1984, in *American Economic Review, Papers and Proceedings*, 74 (May 1984), pp. 67–72.

250   *Bibliography*

Flaherty, M. T. and H. Itami, 'Financial Institutions and Financing for the Semiconductor Race', Graduate School of Business Adminstration, Harvard University, 1984, No. 2.

Flamm, K. and J. Pelzman, *New Forms of Investment by US Firms in Emerging and Declining Sectors: Textiles and Microelectronics*, The Brookings Institution/OECD Research Project on New Forms of Investment in Developing Countries, Paris, OECD, 1984.

Forester, R. (ed.), *The Microelectronics Revolution*, Oxford, Blackwell, 1980.

Fox, M. B., 'The Role of Finance in Industrial Organization: A General Theory and the Case of the Semiconductor Industry', doctoral dissertation, Yale University, 1980.

France, Commissariat Général du Plan, *V Plan 1966-1970: Electronique*, Paris, La Documentation Française, 1966.

— Ministère de l'Industrie et de la Recherche, *Composant électronique*, Paris, La Documentation Française, 1976.

Freeman, C., *The Economics of Industrial Innovation*, London, Frances Pinter and Cambridge, MA, MIT Press, 1982.

—, J. Clark and L. L. G. Soete, *Unemployment and Technical Innovation: A Study of Long Waves in Economic Development*, London, Frances Pinter, 1982.

—, C. J. Harlow and J. K. Fuller, 'Research and Development in Electronic Capital Goods', *National Institute Economic Review*, No. 34 (1965).

Freund, R. E., 'Competition and Innovation in the Transistor Industry', doctoral dissertation, Duke University, 1971.

Fusfeld, H. and J. Tooker, *Status of French and German Electronic Industry*, New York, Center for Science and Technology Policy, 1980.

Futia, C., 'Schumpeterian Competition', *Quarterly Journal of Economics*, **94** (1980), pp. 675-95.

Goetzeler, H., 'Zur Geschichte der Halbleiter', *Technikgeschichte*, **39** (1972), n. 1.

Gold, B., 'Technological Diffusion in Industry: Research Needs and Shortcomings', *Journal of Industrial Economics*, March 1981.

Golding, A. M., 'The Semiconductor Industry in Britain and the United States: A Case Study in Innovation, Growth and the Diffusion of Technology', doctoral dissertation, University of Sussex, 1971.

Gougeon, P., B. Ponson and Y. Tinard, 'La filière électronique', *Regards sur l'Actualité*, No. 100, 1984.

Grabowski, H. G. and J. M. Vernon, 'The Pharmaceutical Industry', in R. Nelson (ed.), *Government and Technical Progress*, New York, Pergamon Press, 1982.

Grant, R. M. and G. K. Shaw, 'Structural Policies in West Germany and the United Kingdom towards the Computer Industry', unpublished paper, New York, 1979.

Hannan, M. T. and C. Freeman, 'The Population Ecology of Organizations', *American Journal of Sociology*, **82** (1977), pp. 929-64.

Harman, A., *The International Computer Industry*, Cambridge, MA, Harvard University Press, 1971.

Harrigan, K. R., *Strategies for Vertical Integration*, Lexington Books, Lexington, MA, 1983.

Hazewindus, N., *The US Microelectronics Industry*, New York, Pergamon Press, 1982.

Heiner, R., 'The Origin of Predictable Behavior', *American Economic Review*, 73, No. 4 (1983).

Hicks, J., *Casuality in Economics*, Oxford, Basil Blackwell, 1979.

High, J., 'Knowledge, Maximizing and Conjecture: a Critical Analysis of Search Theory', *Journal of Post-Keynesian Economics*, Winter 1983-4.

Hill, C. and J. Utterback (eds), *Technological Innovation for a Dynamic Economy*, New York, Pergamon Press, 1979.

Hills, J., *Information Technology and Industrial Policy*, London, Croom Helm, 1984.

Hirsch, W. Z., 'Manufacturing Progress Functions', *Review of Economics and Statistics*, 34 (1952), pp. 143-55.

—, 'Firm Progress Ratios', *Econometrica*, 24 (1956), pp. 136-43.

Hirschman, A., *The Strategy of Economic Development*, Yale University Press, New Haven, 1958.

Hisch, A., 'An International Trade and Investment Theory of the Firm', *Oxford Economic Papers*, 28, July 1976, pp. 258-69.

Hu, Y. S., *The Impact of US Investments in Europe: A Case Study of the Automotive and Computer Industries*, New York, Praeger, 1973.

Hymer, S. H., *The International Operations of National Firms: A Study of Direct Foreign Investment*, Cambridge, MA, MIT Press, 1976.

Ijiri, Y. and H. A. Simon, 'Business Firm Growth and Size', *American Economic Review*, 54 (1964), pp. 77-89.

Integrated Circuit Engineering Corporation, *Status: A Report on the Integrated Circuit Industry*, Scottsdale, ICE, various years.

Italy, Ministero dell'Industria, *Programma finalizzato Elettronica*, Rome, 1979.

—, Ministero della Ricerca Scientifica e Tecnologica, *Programma Nazionale di Ricerca per la Microelettronica*, Rome, 1983.

Jequier, N., 'Computers', in R. Vernon (ed.), *Big Business and the State*, Cambridge, MA, Harvard University Press, 1974.

Journal of Electronic Engineering, VLSI Business Strategies, Part 5, June 1982.

Jublin, J. and J. M. Quatrepoint, *French Ordinateurs*, Paris, Alain Moreau, 1976.

Kamien, M. I. and N. L. Schwartz, *Market Structure and Innovation*, Cambridge, Cambridge University Press, 1981.

Katz, B. and A. Phillips, 'Government, Technological Opportunities and the Structuring of the Computer Industry: 1946-61', in R. R. Nelson (ed.), *Government and Technical Progress: A Cross-Industry Analysis*, New York, Pergamon Press, 1982.

Keck, O. 'West German Science Policy Since the Early 1960s: Trends and Objectives', *Research Policy*, 5 (1976).

—, *Policy-making in a Nuclear Reactor Programme: the Case of the West German Fast Breeder Reactor*, Lexington Books, Lexington, MA, 1981.

Kindleberger, C., *American Business Abroad*, New Haven, CT, Yale University Press, 1969, pp. 1-36.

Kleiman, H., 'The Integrated Circuit: A Case Study of Product Innovation', doctoral dissertation, George Washington University, 1966.

Klein, B. H., *Dynamic Economics*, Cambridge, MA, Harvard University Press, 1977.

Kloten, N., A. Ott, W. Gosele and R. Pfeiffer, *Der EDV-Market in der Bundes Republik Deutschland*, Tübingen, Mohr, 1976.

Knickerbocker, F. T., 'Oligopolistic Reaction and Multinational Enterprise', Graduate School of Business Administration, Harvard University, 1973.

Knight, F., *Risk, Uncertainty and Profit*, Boston, MA, Houghton Mifflin, 1921.

Kraus, J., 'The British Electron Tube and Semiconductor Industry, 1935–62', *Technology and Culture*, 9, No. 4 (1968), pp. 544–61.

——, 'An Economic Study of the US Semiconductor Industry', doctoral dissertation, New School for Social Research, 1973.

Krugman, P., 'The US Response to Foreign Industrial Targeting', *Brookings Papers on Economic Activity*, 1, 1984.

Lake, A., 'Transnational Activity and Market Entry in the Semiconductor Industry', National Bureau of Economic Research Working Paper No. 126, March 1976.

Lamborghini, B. and C. Antonelli, 'The Impact of Electronics on Industrial Structures and Firms' Strategies', in *Microelectronics, Productivity and Employment*, Paris, OECD, 1981, pp. 77–121.

Lancaster, K., 'A New Approach to Consumer Theory', *Journal of Political Economy*, 74 (1966), pp. 132–57.

——, *Variety, Equity and Efficiency*, New York, Columbia University Press, 1979.

Landes, D., *The Unbound Prometheus*, Cambridge, Cambridge University Press, 1969.

Langlois, R., *Internal Organization in a Dynamic Context: Some Theoretical Considerations*, New York, Draft, 1983.

Layton, C., *Ten Innovations*, London, George Allen and Unwin, 1972.

Le Bolloch, C., *L'intervention de l'Etat dans l'Industrie Eléctronique en France de 1974 à 1981*, doctoral dissertation, Université de Rennes I, UER de Sciences Économiques, 1983.

Levin, R. C., 'Innovation in the Semiconductor Industry: Is a Slowdown Imminent?' in H. I. Fusfeld and R. N. Langlois (eds), *Understanding R&D Productivity*, New York, Pergamon Press, 1982.

——, 'The Semiconductor Industry', in R. R. Nelson (ed.), *Government and Technical Progress: A Cross-Industry Analysis*, New York, Pergamon Press, 1982.

Levy, J., 'Diffusion of Technology and Patterns of International Trade: The Case of Television Receivers', doctoral dissertation, Yale University, 1981.

Lewicki, A., *Einführung in die Mikroelektronik*, Munich, R. Oldenburg, 1966.

Linvill, J. G., A. Lamond and R. Wilson, *The Competitive Status of the US Electronics Industry*, Washington, National Academy Press, 1984.

Lizzeri, C. and F. De Brabant, *L'industria delle telecomunicazioni in Italia*, Milan, F. Angeli, 1979.

Lorenzi, J. H. and E. LeBoucher, *Mémoires volées*, Paris, Ramsay, 1979.

Loury, G., 'Market Structure and Innovation', *Quarterly Journal of Economics*, 93 (1979), pp. 395–410.

MacDonald, S., 'Much Ado About Patents', Brisbane, University of Queensland, 1981, mimeo.

McGraw-Hill, *Encyclopedia of Science and Technology*, New York, McGraw-Hill, 1977.

McLean, M., 'The Electronics Industry', in *Technical Change and Economic Policy*, Paris, OECD, 1980.

Mansfield, E., 'Technical Change and the Rate of Imitation', *Econometrica*, 29 No. 4 (1961), pp. 741–66.

—, *Industrial Research and Technological Innovation: An Econometric Analysis*, New York, Norton, 1968.

—, J. Rapoport, A. Romeo, E. Villani, S. Wagner and F. Husic, *The Production and Application of New Industrial Technology*, New York, Norton, 1977.

—, M. Schwarz and S. Wagner, 'Imitation Costs and Patents: an Empirical Study', *Economic Journal*, December 1981.

—, *Technology Transfer, Productivity and Economic Policy*, New York, Norton, 1982.

March, J. G. and H. Simon, *Organizations*, New York, Wiley, 1958.

Mariotti, S. and P. Migliarese, 'Organizzazione industriale e rapporti fra imprese in un settore ad elevato tasso innovativo', *L'Industria*, 5 (1984), pp. 71–110.

Masuda, Y. and W. Steinmuller, 'The Role of Vertical Integration in the US and Japanese Semiconductor Industries', Draft Paper, US–Japan Relations Group, Stanford University, 1981.

Metcalfe, J., 'Industrial Policy and the Evolution of Technology', paper presented at the Conference 'Innovazioni technologiche e struttura produttiva: la posizione dell'Italia', Milan, 1983.

Momigliano, F., 'Piano a medio termine e proposte di rielaborazione degli strumenti di politica industriale in Italia', *L'Industria*, 3 (1981).

Mowery, D., 'The Nature of the Firm and the Organization of Research: An Investigation of the Relationship between Contract and In-house Research', Graduate School of Business Administration, Harvard University, January 1982.

—, 'Innovation, Market Structure and Government Policy in the American Semiconductor Electronics Industry: A Survey', *Research Policy*, 4 (1983).

— and N. Rosenberg, 'The Influence of Market Demand upon Innovation: A Critical Review of Some Recent Empirical Studies', in N. Rosenberg (ed.), *Inside the Black-Box: Technology and Economics*, Cambridge, MA, Harvard University Press, 1982.

NEDO, *The Microelectronics Industry*, London, NEDO, 1980.

—, *Innovation in the UK*, London, NEDO, 1982.

Nelson, R. R., 'The Simple Economics of Basic Scientific Research', *Journal of Political Economy*, 67 (1959), pp. 297–306.

—, 'Uncertainty, Learning and the Economics of Parallel R and D', *Review of Economics and Statistics*, 43 (1961), pp. 351–64.

—, 'The Link between Science and Invention: The Case of the Transistor', in *The Rate and Direction of Inventive Activity—Economic and Social Factors*,

254   *Bibliography*

US Bureau of Economic Research, Princeton, Princeton University Press, 1962, pp. 549–83.

—, 'Issues and Suggestions for the Study of Industrial Organization in a Regime of Rapid Technical Change', in V. R. Fuchs (ed.), *Policy Issues and Research Opportunities in Industrial Organization*, National Bureau of Economic Research, New York, Columbia University Press, 1972.

—, 'Research on Productivity Growth and Productivity Differences: Dead Ends and New Departures', *Journal of Economic Literature*, **19** No. 3 (September 1981), pp. 1029–64.

—, (ed.), *Government and Technical Progress*, Pergamon Press, New York 1982.

—, *High-Technology Policies: A Five-Nation Comparison*, American Enterprise Institute, Washington, 1984.

—, M. J. Peck and E. Kalachek, *Technology, Economic Growth and Public Policy*, Washington, DC, The Brookings Institution, 1967.

Nelson, R. and S. Winter, 'Neoclassical vs. Evolutionary Theories of Economic Growth: Critique and Prospectus', *Economic Journal*, December, 1974.

— and —, *An Evolutionary Theory of Economic Change*, Cambridge, Belknap Press, 1982.

Newfarmer, R., *The International Market Power of Transnational Corporations: A Case Study of the Electrical Industry*, UNCTAD, New York, United Nations, 1978.

Nomura, Research Department, *Microchip Revolution in Japan*, Tokyo, 1980.

Noyce, R. N., 'Microelectronics', in T. Forester (ed.), *The Microelectronics Revolution*, Cambridge, MA, MIT Press, 1980.

OECD, 'Gaps in Technology-Electronic Components', Paris, 1968.

—, 'Gaps in Technology-Electronic Computers', Paris, 1969.

—, 'Impact des entreprises multinationales sur les potentiels scientifiques et techniques nationaux. Industrie des ordinateurs et de l'informatique.' Prepared by A. Michalet and M. Delapierre, Paris, 1977.

—, 'Telecommunication Equipment Industry Study', Paris, 1981.

Pavitt, K., *Technology Transfer among the Industrially Advanced Countries*, Paper delivered at a conference held in New York City, 2–3 June 1983.

—. *Technical Innovation and British Economic Performance*, London, Macmillan, 1980.

— and W. Walker, 'Government Policies Towards Industrial Innovation: A Review', *Research Policy*, **5** No. 1 (1976), pp. 11–97.

Peck, M. J., 'Inventions in the Postwar American Aluminium Industry', in *The Rate and Direction of Inventive Activity*, Princeton, Princeton University Press, 1962, pp. 279–98.

—, 'Government Coordination of R-D in the Japanese Electronic Industry', in *Dynamic Competition: A Program of Research on Technological Change and Market Structure*, Second Year Progress Report to the National Science Foundation,Department of Economics, Yale University, May 1983,pp. 261–92.

— and Goto, 'Technology and Economic Growth: The Case of Japan', *Research Policy*, **10** No. 3 (1981), pp. 222–43.

— and R. Wilson, 'Innovation, Imitation and Comparative Advantage: The Case of the Consumer Electronics Industry', in H. Giersch (ed.), *Proceedings of Conference on Emerging Technology at Kiel Institute of World Economics*, Tübingen: J. C. B. Mohr, 1981.

Pelc, K. I., 'Remarks on the Formulation of Technology Strategy', in D. Sahal (ed.), *Research, Development and Technological Innovation*, Lexington, MA, Lexington Books, 1980.

Perry, M. K. and R. H. Groff, Vertical Integration and Growth: an Examination of the Stigler Story—Bell Laboratories, *Economic Discussion Papers*, 1982.

Pertile, R., *L'industria dell'informatica e dei componenti elettronici: analisi delle prospettive e proposte di intervento*, Rome, ISPE, 1975.

Philips, A., *Technology and Market Structure: A Study of the Aircraft Industry*, Lexington, MA, D. C. Heath, 1971.

—, 'Organizational Factors in R and D and Technological Change: Market Failure Considerations', in D. Sahal (ed.), *Research, Development and Technological Innovation*, Lexington, MA, Lexington Books, 1980.

Porter, M. E., 'The Structure Within Industries and Companies' Performance', *Review of Economics and Statistics*, 61 No. 2 (May 1979), pp. 214–27.

Porter, M., *Competitive Strategy*, Free Press, New York, 1980.

—, 'The Technological Dimension of Competitive Strategy' in *Research on Technological Innovation, Management and Policy*, 1, pp. 1–33, JAI Press, Greenwich, 1983.

Pottier, C. and P. Y. Touati, 'Concurrence internationale et localisation de l'industrie des semi-conducteurs en France', *EEE*, No. 24, Paris, Université de Paris I, Panthéon, Sorbonne, 1981.

Pugel, T. A., Y. Kimara and R. G. Hawkins, 'Semiconductors and Computers: Emerging International Competitive Battlegrounds', in R. W. Moxen, R. W. Roehl and J. S. Truitt (eds), *International Business Strategies in the Asia-Pacific Region*, JAI Press, Greenwich (CT), 1983.

Quatrepoint, J. M. and J. Jublin, *French Ordinateurs*, Paris, A. Moreau, 1976.

Reinganum, J., 'A Dynamic Game of R and D: Patent Protection and Competitive Behavior', *Econometrica*, 50, No. 3, May 1982, pp. 671–88.

Rosenberg, N., *Technology and American Economic Growth*, New York, 1972.

—, 'Perspective on Technology', Cambridge, MA Cambridge University Press, 1976.

—, *Inside the Black Box: Technology and Economics*, Cambridge, MA, Cambridge University Press, 1982.

— and C. Fritschak, 'Long Waves and Economic Growth: A Critical Appraisal', *American Economic Review Papers and Proceedings*, May 1983, pp. 146–51.

— and E. Steinmuller, 'The Economic Implications of the VLSI Revolution', *Futures*, October 1980.

Rosenbloom, R. and W. Abernathy, 'The Climate for Innovation in Industry', *Research Policy*, II, 1982, pp. 209–25.

Rösner, A., *Die Wettbewerbverhältnisse auf der Markt für elektronische Datenverarbeitungsanlagen in der BRD*, Berlin, Duncker-Humblot, 1978.

Rugman, A. M., *Inside the Multinationals: The Economics of Internal Markets*, New York, Columbia University Press, 1981.

Sahal, D., *Patterns of Technological Innovation*, Reading, MA, Addison-Wesley, 1981.

Sampson, A., *The Sovereign State of ITT*, New York, Stein and Day, 1973.

Saviotti, P. P. and J. S. Metcalfe, 'A Theoretical Approach to the Construction of Technological Output Indicators', *Research Policy*, **13**, No. 3, June 1984, pp. 141–52.

Scherer, F., 'Inter-industry Technology Flows in the United States', *Research Policy*, **11** No. 4 (1982), pp. 227–46.

Schmalensee, R., 'A Note on the Theory of Vertical Integration', *Journal of Political Economy*, **81**, March–April 1973, pp. 442–9.

Schmookler, J., *Invention and Economic Growth*, Cambridge, MA, Harvard University Press, 1966.

Schnee, J., 'Government Programs and the Growth of High-Technology Industries', *Research Policy*, **7**, 1978, pp. 2–24.

Scholz, L., *Die spezifischen Probleme der Marktforschung und Unternehmensplanung in der Bauelemente-Industrie*, Munich, IFO, 1971.

—, *Technologie und Innovation in der industriellen Produktion*, University of Munich Inaugural Dissertation, 1974.

Schumpeter, J., *The Theory of Economic Development*, Cambridge, MA, Harvard University Press, 1934.

—, *Business Cycles: A Theoretical, Historical and Statistical Analysis of the Capitalist Process*, New York and London, McGraw-Hill, 1939.

—, *Capitalism, Socialism, Democracy*, New York, Harper, 1950.

—, *Essays*, R. Clemence (ed.), Cambridge, MA, Addison-Wesley Press, 1951.

Schwartzman, D., *Innovation in the Pharmaceutical Industry*, John Hopkins University Press, Baltimore, 1976.

Sciberras, E., *Multinational Electronics Companies and National Economic Policies*, Greenwich, CT, JAI Press, 1977.

—, 'The U.K. Semiconductor Industry', in K. Pavitt (ed.), *Technical Innovation and British Economic Performance*, London, Macmillan, 1980.

—, N. Swords-Isherwood and P. Senker, 'Competition, Technical Change and Manpower in Electronic Capital Equipment: A Study of the U.K. Minicomputer Industry', *SPRU Occasional Paper No. 8*, University of Sussex, 1978.

Semiconductor Industry Association, *The International Microelectronic Challenge*, Cupertino, CA, Semiconductor Industry Association, May 1981.

Shackle, G. L. S., *Epistemics and Economics*, Cambridge University Press, Cambridge, 1972.

Simon, H., *Models of Man*, New York, John Wiley, 1957.

—, 'Theories of Bounded Rationality', in B. McGuire and R. Radner (eds), *Decision and Organization*, North Holland, New York, 1972.

—, 'Rational Decision Making in Business Organizations', *American Economic Review*, **69**, No. 4 September 1979, pp. 493–513.

—, 'On the Behavioral and Rational Foundations of Economic Dynamics', *Economic Review*, **69**, No. 4 (September 1979, pp. 493–513.

Soria, L., *Informatica: un'occasione perduta*, Turin, Einaudi, 1979.

Stigler, G. J., 'The Division of Labor Is Limited by the Extent of the Market',

Journal of Political Economy, **59**, June 1951, pp. 185–93.

Stoke, R., 'Government Electronics: Federal Output Tough for Foreigners', *Electronics*, **41**, 9 December 1968, pp. 119–24.

Stoneman, P., *The Economic Analysis of Technological Change*, Oxford University Press, New York, 1983.

Streetman, B. *Solid State Electronic Devices*, New York, Prentice Hall, 1980.

Stuckey, J. A., *Vertical Integration and Joint Ventures in the Aluminium Industry*, Harvard University Press, Cambridge, 1983.

Suleiman, E., 'Industrial Policy Formulation in France', in S. Warnecke and E. Suleiman (ed.), *Industrial Policies in Western Europe*, New York, Praeger, 1975.

Symposium on Uncertainty (articles by R. Bansor, M. Rutherford, G. L. S. Shackle, B. J. Loasby, and D. L. Weisman), *Journal of Post Keynesian Economics*, **VI**, No. 3, Spring 1984, pp. 340–420.

Taylor, C. T. and Z. A. Silberston, *The Economic Impact of the Patent System*, Cambridge, UK, Cambridge University Press, 1973.

Teece, D., 'Technology Transfer by Multinational Firms: The Research Cost of Transferring Technological Know-How', *Economic Journal*, **87**, June 1977, pp. 242–61.

—, 'Economies of Scope and the Scope of the Enterprise: The Diversification of Petroleum Companies', *Journal of Economic Behavior and Organization*, **1**, 1980, pp. 223–47.

—, 'The Multinational Enterprise: Market Failure and Market Power Considerations', *Sloan Management Review*, Sping 1981.

—, 'Towards an Economic Theory of the Multiproduct Firm', *Journal of Economic Behavior and Organization*, **3**, 1982, pp. 39–64.

Teubal, M., 'On User Needs and Need Determinations: Aspects of the Theory of Technological Innovation', in *Industrial Innovation*, M. Baker (ed.), London, Macmillan, 1979.

—, 'The Accumulation of Intangibles by High-Technology Firms' in *The Trouble with Technology*, S. MacDonald, D. McL. Lamberton and T. D. Mandeville (eds), London, Frances Pinter, 1983.

Tilton, J., *International Diffusion of Technology: The Case of Semiconductors*, Washington, DC, The Brookings Institution, 1971.

Truel, J. L., 'L'Industrie Mondiale des Semi-Conducteurs', doctoral dissertation, Université de Paris-Dauphiné, 1980.

—, 'Perspective de l'industrie française des semi-conducteurs', unpublished paper, Paris, 1981.

—, 'Structuration en filière et politique industrielle dans l'électronique: une comparaison internationale', *Revue d'Economie Industrielle*, No. 23, I° trim., 1983, pp. 293–303.

Uenohara, M. *et al.*, *Constrasting Patterns of Technological Development*, Draft Report on US–Japanese Semiconductor Industry Competition, Forum on International Policy, Stanford University, 1982.

UNCTAD, *International Subcontracting Arrangements in Electronics between Developed Market-Economy Countries and Developing Countries*, New York, United Nations, 1975.

UK Department of Trade and Industry, *A Programme for Advanced Information Technology: the Report of the Alvey Committee*, London, 1983.

UK Electronic Components Economic Development Committee, *The Electronic Component Industry*, London, March 1983.

United Nations—Centre on Transnational Corporations, *Transnational Corporations in the International Semiconductor Industry*, New York, 1983.

United States Bureau of Economic Research, *The Rate and Direction of Inventive Activity—Economic and Social Factors*, Princeton, Princeton University Press, 1962.

—, Congress—Office of Technology Assessment, *International Competitiveness in Electronics*, Washington, 1983.

—, Department of Commerce, *The American Computer Industry in Its International Competitive Environment*, Washington, DC, US Government Printing Office, 1976.

—, Department of Commerce, Business and Defense Services Adminstration, *Electronic Components: Production and Related Data, 1952–1959*, Washington, DC, Department of Commerce, 1960.

—, *Report on the Semiconductor Industry*, Washington, DC, US Government Printing Office, 1979.

—, Trade Commission, *Competitive Factors Influencing World Trade in Integrated Circuits*, Washington, DC, US Government Printing Office, 1979.

Usher, A. P., *A History of Mechanical Inventions*, Cambridge, Harvard University Press, 1954.

Utterback, J., 'The Dynamics of Product and Process Innovation in Industry', in C. Hill and J. Utterback, *Technological Innovation for a Dynamic Economy*, New York, Pergamon Press, 1979.

— and W. Abernathy, 'A Dynamic Model of Process and Product Innovation', *Omega*, **3**, No. 6, 1975.

Utterback, J. M. and A. Murray, *Influence of Defense Procurement and Sponsorship of Research and Development on the Civilian Electronics Industry*, Cambridge, MA, MIT Center for Policy Alternatives, 1977.

Vernon, J. and D. Graham, 'Profitability of Monopolization by Vertical Integration', *Journal of Political Economy*, **79**, July–August 1971, pp. 924–5.

Von Hippel, E., 'The Dominant Role of Users in Scientific Instruments Innovation Process', *Research Policy*, **5** No. 3, 1976, pp. 212–39.

—, 'The Dominant Role of Users in Semiconductors and Electronic Subassembly Process Innovation', *IEEE Transactions on Engineering Management*, EM **24**, May 1977, pp. 60–71.

—, 'A Customer-Active Paradigm for Industrial Product Idea Generation', *Research Policy*, **7** No. 2, 1978, pp. 240–66.

—, 'Appropriability of Innovation Benefit as a Predictor of the Source of Innovation', *Research Policy*, **11** No. 2, 1982, pp. 95–116.

Von Weiher, S. and H. Goetzeler, *The Siemens Company: Its Historical Role in the Progress of Electrical Engineering*, Munich, Bruckmann, 1984.

Warren-Boulton, F., 'Vertical Control with Variable Proportions', *Journal of Political Economy*, **82** No. 4, July–August 1974, pp. 783–802.

Webbink, D. A., *The Semiconductor Industry: A Survey of Structure, Conduct and Performance*, Staff Report to the Federal Trade Commission, Washington, DC, Federal Trade Commission, 1977.

Williamson, O., *Corporate Control and Business Behavior*, Englewood Cliffs, NJ, Prentice-Hall, 1970.

—, *Markets and Hierarchies: Analysis and Antitrust Implications*, New York, Macmillan, 1975.

—, 'Transaction-Cost Economics: The Governance of Contractual Relations', *Journal of Law and Economics*, **22**, 1979, pp. 233–61.

—, 'The Modern Corporation: Origin, Evolution, Attributes', *Journal of Economic Literature*, **19** No. 4, December 1981, pp. 1537–70.

Wilson, R., P. Ashton and T. Egan, *Innovation, Competition and Government Policy in the Semiconductor Industry*, Cambridge, MA, Lexington Books, 1980.

Young, E. C., *The New Penguin Dictionary of Electronics*, London, Penguin, 1979.

Zysman, J., *Political Strategies for Industrial Order*, Berkeley, CA, University of California Press, 1977.

—, 'Between the Market and the State: Dilemmas of French Policy for the Electronics Industry', *Research Policy*, **3**, 1975.

—, *Governments, Markets and Growth: Financial Systems and the Politics of Industrial Change*, Cornell University Press, Ithaca, 1983.

— and L. Tyson, *American Industry in International Competition: Government Policies and Corporate Strategies*, Cornell University Press, Ithaca, 1983.

Index